ART FOR PSYCHOTHERAPY

ART FOR PSYCHOTHERAPY

ISBN: **978-1-326-78959-6**

Andreas Sofroniou   2016 © Copyright

All rights reserved.

Making unauthorised copies is prohibited.

No parts of this publication may be reproduced, transmitted, transcribed, stored in a retrieval system, translated in any language, or computer language, in any form, or by any means, without the prior written permission of Andreas Sofroniou.

Andreas Sofroniou   2016 © Copyright

ART FOR PSYCHOTHERAPY

ISBN: **978-1-326-78959-6**

ART FOR PSYCHOTHERAPY

CONTENTS   PAGE

SECTION ONE:
EXPRESSION OF ART - SCRIBBLING, DOODLING, SKETCHING, SCULPTURE 18

SECTION TWO:
CONDENSED SUMMARY OF REPORT ON UNDERSTANDING CHILDREN'S ART 20

SECTION THREE:
ART FORMS FOR DIADNOSTIC METHODS USED 26
Doodles 26
Sketch 26
Drawing 28
Elements and principles of design 30
Plane techniques 34
Drawing surface 35
Relationship between drawing and other art forms 36
Surfaces 37
Tools and techniques 39
Charcoal 40
Chalks 42
Metalpoints 47
Graphite point 48
Coloured crayons 50
Incised drawing 50

ART FOR PSYCHOTHERAPY

Brush, pen, and dyestuffs	51
Pens	52
Inks	53
Pen drawings	55
Brush drawings	59
Combinations of various techniques	62
Mechanical devices	66
Applied drawings	68
Subject matter of drawing	71
Portraits	72
Landscapes	74
Figure compositions and still lifes	77
Fanciful and non-representational drawings	78
Artistic architectural drawings	79
History of drawing – Western	80
Centuries: 14th, 15th, and 16th	80
Centuries: 17th, 18th, and 19th	84
Modern	89
Eastern	90
SECTION FOUR: SCULPTURE	92
Sculpture as an art	92
Elements and principles of sculptural design	95
Elements of design	96
Principles of design	99

ART FOR PSYCHOTHERAPY

Relationships to other arts	104
Materials	105
Primary	105
Secondary	114
Methods and techniques	116
Sculptor as designer and as craftsman	117
General methods	119
Carving	121
Indirect carving	122
Carving tools and techniques	123
Modelling	124
Modelling for casting	125
Modelling for pottery sculpture	127
General characteristics of modelled sculpture	127
Constructing and assembling	128
Direct metal sculpture	130
Reproduction and surface-finishing techniques	131
Casting and moulding	131
Pointing	135
Surface finishing	136
Smoothing and polishing	136
Painting	137
Gilding	137
Patination	138
Electroplating	138

ART FOR PSYCHOTHERAPY

Other finishes	139
Symbolism of sculpture	139
Sculpture in the round	139
Relief sculpture	142
Modern forms of sculpture	144
Representational sculpture	147
Human figure	148
Devotional images and narrative sculpture	149
Portraiture	150
Scenes of everyday life	150
Animals	151
Fantasy	152
Other subjects	152
Non-representational sculpture	153
Decorative sculpture	155
Symbolism	156
Uses of sculpture	159
SECTION FIVE: **SKETCHES & DRAWINGS CONCEPTS**	162
Schematic comparisons	162
Interpreting Freudian Concepts	163
Conscious and Subconscious	163
Copernicus of the Mind	164
Hypnosis	164

ART FOR PSYCHOTHERAPY

Repression	**165**
Impulses	**165**
Repressions	**165**
Conscious, Ego, Sub-conscious	**166**
Psychoanalytical Society	**166**
Jung and Adler	**167**
Breaking up	**167**
Hitler	**167**
Royal Society	**168**
Writings	**168**
Inter-relationship of Sciences	**168**
Complex	**169**
Listening	**169**
London Work	**170**
Methodology	**170**
Doctrines	**170**
Observations and Experiences	**171**
Oedipus	**171**
Philosophical Thought	**171**
Psychoanalysis in the Sphere of Philosophy	**171**
Unknown Points	**172**
Hypotheses	**172**
Individual Development	**172**
Conscious, Preconscious, Ego, and Subconscious	**173**
Mental Regions	**173**

ART FOR PSYCHOTHERAPY

Ego	**174**
Id	**174**
Tensions	**175**
Dependence	**175**
Id, Super-ego, and Reality	**176**
Substitutes	**176**
Physical Apparatus	**176**
Region of Complex	**177**
Theory of the Instincts	**178**
Needs of the Id	**178**
Displacement	**178**
Self-preservation	**179**
Eros	**179**
Opposing Forces	**180**
Sexual Aggressiveness	**180**
Restrictions	**180**
Libido	**181**
Psychê	**181**
Diversion	**182**
Fixation	**182**
Self-destructiveness	**182**
Internal Conflicts	**183**
Narcissistic	**183**
Libidinal Cathexes	**183**
Erotogenic Zones	**184**

ART FOR PSYCHOTHERAPY

Eros	**184**
Sexual Impulse	**184**
Super-ego, Ego, Instincts	**185**
Development of the Sexual Function	**185**
Fruitful Relationship	**186**
Attractions	**187**
Sexuality	**187**
Bodily Activity	**188**
Phenomena	**188**
Infantile Amnesia	**188**
Aetiology	**189**
Libidinal Demands	**189**
Obstinate Persistence	**189**
Sadistic Impulses	**189**
Phallic Phase	**190**
Penis Presence	**190**
Oedipus Phase	**190**
Electra Phase	**191**
Simultaneous Presence	**191**
Independent Pleasure	**191**
Puberty	**192**
Inhibitions	**192**
Homosexuality	**192**
Cathexes	**193**
Tendency	**193**

ART FOR PSYCHOTHERAPY

Phenomenology	193
Dynamics and Economics	193
Aetiology of the Disturbances	194
Antithesis	194
Mental Structure	195
Mental Life	195
Consciousness	195
Consciousness Assumption	195
Conscious Processes	196
Somatic Processes	196
Fundamental Hypothesis	196
Psycho-analysis and Philosophy Disputes	197
Natural Science	197
Apparatus	197
Conscious Events	197
Mental Processes	198
Internal Resistances	198
Perceptions	198
Peremptory Influence	199
Energy	199
Cathexes and Hyper-cathexes	199
Primary Process	199
Dream Interpretation	200
Conflict and Rebellion	200
Dreams Normality	200

ART FOR PSYCHOTHERAPY

Dreams	201
Dream-process	201
Dream Manifestation	201
Ego Forcing	201
Mechanism of Dreams	202
Superfluity	202
Evidence of Dream-work	202
Subconscious Work-over	203
Kingdom of the Illogical	203
Compromise	203
Wish Fulfilment	204
Product of Conflict	204
Subconscious Mechanism	204
Super-ego Claims	205
Disturbed Reality	205
Jungian Syntheses	205
Studies and Experience	206
Critical of Freud	206
Self –regulating System	206
Symbolism	207
Influences on Sciences	207
Condensed Picture	207
Short Task	207
Science not Philosophy	208
Term Established	208

ART FOR PSYCHOTHERAPY

Philosophical Derivation	208
Psychic Equals Physical	208
Interchangeable Subjects	209
Psychological Aspect	209
Principles	210
Psyche, Soul, or Mind	210
Ego	210
Centre of Reference	211
Consciousness	211
Personal Unconscious	212
Primal Datum	212
Unconscious Sphere	213
Consciousness Dependent on Unconscious	213
Unconscious Contents	214
Psychic Functions	215
Decisive Facts	216
Rational Functions	216
Irrational Functions	216
Inner Perception	217
Historical Event	217
Dominant Function	217
Typology	217
Mixed Types	218
Over-differentiation	219
Persona	219

ART FOR PSYCHOTHERAPY

Sphere of Consciousness	**219**
Function-complex	**220**
Stigmatisation	**221**
Personality Inflation	**221**
Psychic Health	**221**
Character Index	**222**
Extraversion and Introversion	**222**
Habitus, the Central Switchboard	**222**
Unsuccessful Adjustment	**222**
Orientation of Value	**223**
Empirical Material	**223**
Conscious Effort	**223**
Inner Rebuilding	**224**
Ancillary Functions	**224**
Compensatory Relation	**224**
Matrimonial Problems	**224**
Eight Different Psychological Types	**225**
Compass	**225**
Sphere of Unconscious	**227**
Regional Divisions	**228**
Autonomous Unconscious	**228**
Zones	**228**
Symptom and Complex	**229**
Broadening of Consciousness	**229**
Arbitrary Functioning	**229**

ART FOR PSYCHOTHERAPY

Nuclear Element	**229**
Ascending Complex	**230**
Passive State	**231**
Moral Conflict	**231**
Complexes	**231**
Psychic Points	**232**
Emotional Shock	**232**
Association Method	**232**
Psyche Mechanism	**232**
Didactic and Diagnostic Method	**233**
Dream	**233**
Interpretation	**233**
Royal Pathway	**233**
Continuity of Processes	**234**
Standard Symbols	**234**
Manifold Contents	**234**
Prophetic	**235**
Fantasies and Visions	**235**
Universal Human History	**235**
Archetypes	**236**
Conscious Adjustment	**236**
Augustine's Term	**236**
Bipolar Structure	**236**
Gestalt	**237**
Axial System	**237**

ART FOR PSYCHOTHERAPY

Clarification	237
Heavy Burden	239
Creative Union	240
Psychic System	241
Jungian System: Neither Religion nor Philosophy	242
Equipment	242
Reassurance and Comfort	242
Reason over Instinctive Nature	243
Transformation	243
Culture of the Future	243
Psychotherapies	243
Psychiatry	244
Child Development	246
Child Psychology	248
Moral Standards	249
SECTION SIX: CINEMATOGRAPHY &TECHNOLOGY	251
Motion pictures	251
History	251
Introduction of sound	256
Introduction of colour	260
Wide-screen and stereoscopic pictures	265
Professional motion-picture production	268
Principal parts	268
Camera supports	275

ART FOR PSYCHOTHERAPY

Film	276
Lighting	277
Light sources	277
Light measurement	280
Film processing and printing	282
Sound-recording techniques	256
Optical recording	286
Magnetic recording	287
Double-system recording	288
Sound recordist	289
Microphones	290
Newer techniques	291
SECTION SEVEN: ART THERAPY & CONCEPTS	293
Memory	293
Freud, Sigmund (1856-1939),	294
Psychotropic drug	295
Tranquillizer	297
Remembering	297
Time-dependent aspects of memory	300
Working memory	300
Executive attention	301
Patterns of acquisition in working memory	302
Long-term memory	304
Patterns of acquisition in long-term memory	305

ART FOR PSYCHOTHERAPY

Intervals	*306*
Rehearsal	*307*
Mnemonic systems	*307*
Physiological aspects of long-term memory	309
Retrieval	312
Relearning	315
Autobiographical memory	316
Eyewitness memory	317
Forgetting	318
Interference	319
Challenges to interference theory	322
Amnesia	323
Aging	325
Causes of Amnesia	327
SECTION EIGHT: SORCERY OR MAGIC	328
Magic	328
Nature and scope	328
Elements of magic	329
Material	330
Rites and condition of the performer	331
Functions	332
Definitional issues: magic, religion, and science	333
Magic and religion	334
Magic and science	335

ART FOR PSYCHOTHERAPY

Subcategories of magic	336
Conceptual history	337
Ancient Mediterranean world	338
Medieval Europe	340
Late medieval and early modern Europe	342
European traditions and the modern world	344
Globalization of the magic concept	346
World cultures	347
Post-colonial points of views	350
History of magic theories	350
Sociological theories	351
Psychological theories	352
Comparative religions	353
Post-modern dialogue	354
Sorcery	355
Conclusion on magic and sorcery	356
SECTION NINE: PROFESSIONAL ART THERAPY	357
Skilled communication for therapy	357
State of consciousness	358
Formulation of a session	359
Ethical approach	359
INDEX	361
BIBLIOGRAPHY	377

ART FOR PSYCHOTHERAPY

SECTION ONE: EXPRESSION OF ART - SCRIBBLING, DOODLING, SKETCHING, SCULPTURE

In 1966 to 1968 I followed a series of studies in England under the headings of Advanced Psychological Topics with the Department of Psychology at Bristol University, Mental Health/Mental Illness with the Bristol Royal Infirmary, and with the West of England College of Art where I contacted research into the interpretation of Children's Art.

By this time, I already had four years of experience following my graduation as a Doctor of Psychology. My research and major studies pertained to mental illnesses and abnormal psychology which took place at the Surrey-based Cane Hill Psychiatric Hospital and the St. Olave's Day Hospital in East London.

The West of England College of Art research on Children's Art was, therefore, of special interest to me. This course of studies included the diagnostic methodology for children's behaviour and therapy, based on the doodling, scribbling, painting and the overall art of children in the age group of three years and up to early adolescence.

In addition, the research at the West of England College of Art was headed by the famous Swiss Professor R M Rossetti, who in my opinion was far ahead of his time in this field, which was based on the children's actual work-related art and expression of their personal experiences

ART FOR PSYCHOTHERAPY

and inner disturbed feelings. At the time, a team report was published for teachers and child psychologists. Thereafter, various publications followed in international periodicals.

Since the Bristol studies almost half a century passed and as a psychotherapist I still use the principles for the diagnosis of human problems based on the three courses I followed in the sixties. Further more, I expanded the artistic expressions of people in interpreting their dreams and parapraxias.

In this book, I am introducing the concepts of modern art and psychology as far as their principles were accepted at the time and to the present day. I am, also, attempting to summarise the studies dealing with art in general, the doodling, sketching, the scribbling, and their influences in modern counselling and therapy, no matter what the age group is, and the kind of school of ideas the reader follows.

The initial short summary of the results (for research funds) was submitted in January 1967, under the title of Understanding Children's Art. The main points of the précis of this report have been scanned and are shown in the following six pages.

ART FOR PSYCHOTHERAPY

SECTION TWO: CONDENSED SUMMARY OF REPORT

UNDERSTANDING CHILDREN'S ART

During the latter part of Sigmund Freud's career, psycho-analysts implied that psycho-analysis can influence the other human sciences. Freud begun with the observation of hysteria and the other neuroses and gradually extended his investigations to normal psychology, sociology and group psychology. Since his initial investigations as a neurologists on neurosis and until his death in 1939, we can see the steady progress of this 'infant' science.

Consultations with analysts and other counselling professions are now the usual thing in all walks of life. In industry, education, vocation, advertising, politics and religion.

Psycho-analysis has helped to understand the life-old subject of art and in expressing uninhibitedly one's own thoughts. In psychiatric hospitals and in clinics, art is now used for therapeutic purposes and educationists can, through art, easily comprehend their pupils and their unconscious motivations.

Art and psycho-analysis are indeed inseparable. An experienced artist can understand the symbolism of dreams more coherently, than a beginner therapist. Psycho-analysis is now reckoned, by many prominent scientists, as being an art of its own and that, a psycho-analyst must have the gift of an artist to understand a subject as controversial and as dark as the subconscious mind itself.

In western countries, art-teachers are now specially trained to do just this. Their aim is to understand children through art and their participation is minimal. They let the children express their own emotions in their own way and thus, children expressing their innate thoughts gain more understanding, insight, confidence, overcome fears and gain maturity more steadily.

Even before psycho-analysis, art-teachers such as Rousseau, Pestaloggi, Froebel and Spencer started using this method, long before modern orthodox psychologists. Each one of them participated in developing the art of children, as it is today.

ART FOR PSYCHOTHERAPY

Franz Ciseck is probably the most prominent psychological art-teacher. He founded a juvenile art class in Vienna in 1897. His programme was to let the children grow, develop and mature. He had no inspector, no curriculum, no time-table and no supervisors. The children's work (art) was rich, decorative and colourful. Once he was asked how he did it. Ciseck astonished them by saying, 'but I don't do it, I take off the lid and other art-masters drop the lid on'. With love and encouragement Cisek maintained that children have three important reasons for their art creation:

1. The creative instinct,

2. The instinct to set in order and

3. The imitative instinct.

Many believe that art must not be tought over the age of fifteen years. Like psychometrists, they believe that intelligence does not grow over the mental age of fifteen, or sixteen. This way, they maintain, individuals can develop their own style.

Let us now see the professional unique practical use of both, art and psycho-analysis. Many analysts today, watch out for doodles, or scribbles as through these, unawarely, we express our emotions.

The Jungian psychotherapists define two main personality types. The introverts and extraverts. Lowenfeld, the art-teacher describes art as visual and hoptic and other schools of ideas describe them as subjective and objective. During their development of personality, children undergo different stages of their art. Some stages are:

1. The stage of scattered features,

2. Assembly (non-tachtically),

3. Encirclement of features and pre-schematic sketches,

4. Schematic.

Children undergoing their development through these stages should be encouraged to print with both hands at the same time. Grozinger encouraged just this, knowing that children can use both hands and his experience tells us that both lobes of the brain can be helped to develop their motor nerves, better.

We must be cautious in asking children what they have scribbled. If we have to ask them, it must be for the sake of showing interest and it must not be a mere 'what is it?', but ask them to tell us about their creation.

ART FOR PSYCHOTHERAPY

During the first stage of their artistic development, children may draw circles one over the other in one pull, so wanting to express their floating or 'here-I-am' attitude. Their cross vertical and horizontal straight lines denote 'standing' and if our little chap, or young lady, comes with joint vertical lines, this means that she/he feels like walking, probably wandering.

How old are children when they start their artistic creation? No child is the same. Just give them lots of paper and pencils, paint, even charcoal and let them get on with it. This is probably the best way to see your doors and those walls clean again. Children merely make use of the walls because their good parent never thought of supplying them with papers.

Many parents will wonder whether their children are good enough at their particular age. Not all children are alike. A rough time-table, though, will probably help the parents, but remember, children are not machines and must never depend on a time-table entirely.

1st year:	More or less random lines, imitative,
1 year 6 months:	Swinging scribbles with right and left hand, strewn and isolated scribbles (zigzags, spirals atc.),
2 years:	Orderly distribution on paper.
3 years:	Sign phase (schematic), interpreted scribbles (mummy, daddy, bow-wow), cephalopods,
3 years 6 months:	Scenes, drawing of letters,
4 years:	Ornamental additions, differentiations (heads, borders, hair, teeth etc.),
4 years 6 months:	Houses, objects (table, chair, train, motor car),
5 years:	Composition of letters to their nannies,
6 years:	First touch of realism. Urge towards synthesis, more guiding concepts. The so-called childish mistakes (mixed profile, x-ray pictures, lack of orientation, mirror-writing) alphabetic writing, plants (tree, flowers), landscape (sun, sky), animals.
7 years:	Differentiation tending towards orientation (special arrangement, representation of space), proportions, movement, colour.
	Further on: automatism, solidification into schematism, petrification into 'cliche'.

ART FOR PSYCHOTHERAPY

Above all, remember that every child is an artist. The art of your child may be wish-fulfilling, a measure of sympathy, a symbol of fertility, a relief of fear and stress, or it may be of an organising nature.

What should a parent do to help his/her child in expressing emotions and in gaining insight? There are no fixed rules to follow, although Grozinger's ten commandments for the parents of children who do pictures, may assist:

1. Do not regard your child as a Rembrandt, or a Picasso, but as a child,

2. When your child scrawls on the table and walls do not scold it, but take yourself to task. The child has done so, only because you have not kept it supplied with paper.

3. Do not go and ask 'what is this?' the moment your child shows you something it has scribbled, or drawn, or painted.

4. Do not show your child how to do it. Do not rub its nose in nature.

5. See that there is a variety of techniques and materials available.

6. Do not panic if your child seems to be regressing.
Each progressive stage begins with a crisis.

7. Be glad of any progress, even if it leads away from something you, yourself like.

8. Do not be a hypochondriac. Children can stand adults' inevitable trash, better than adults can.

9. Learn to wait and be patient.

10. Be prepared for anything.

Psycho-analysts may interpret dreams, uncover slips of tongue (or any other form of parapraxia), or through symbols bring into light that repressed idea. An experienced art-teacher has similar ways of understanding children through their own scribbles and paintings.

A child who draws vertical lines, tells us how strong it is and the child who draws little circles shows his/her submissive behaviour. If your child has beeen told off a lot lately, then watch out for smeared vertical lines, cut-off lines, or crossed vertical lines. The child is probably resisting its own aggression, or even has the sexual drives repressed.

Other tendencies of children, like writing on top of the page, are on top form with their mental and physical aspirations. The opposite of this, can be seen in children who write at the bottom of the page.

ART FOR PSYCHOTHERAPY

Playing with clay and mud is a messy business for the parent. But have we ever thought of the value of this expressiveness, of the dreams a child builds, of her/his free visualisation, or come to think of it - his/her future. What an architect this young person may one day be. Just let the child interact in his/her own little creations, freely.

In children using paints, it is easier to see the state of their minds, their moods and feelings. The red colour is for sociability and affection, yellow for their desire to regress - to remain infatile. With blue, children express their desires and with brown the wish to smear. Green - neutral or unemotional, black - lack of free emotional float.

Painting one colour over another one, means they are hiding their emotions.

Whatever children do, do not forget that their creation through art is an active exploration and an education of consciousness. Art is intelligence and above all, art is feeling.

In adult behaviour, paintings can be used to diagnose behavioural problems, to see the progress an individual makes, or whether regressing. But, remember painting must not be used for cure.

In different counselling sessions, one finds various types of paintings. An individual in a state of depression will not want to paint, or the expression on canvas will be one with an empty horizon and with dark colours. The drawings will be repetively the same.

A person in an over-active and high-spirited frame of mind will produce very exciting paintings with bright colours.

Those in a state of confusion do not produce much.

People with flighty ideas will present their thoughts dissociately and in deep dark colours, or without a connected space or theme. We, also, find that their work is delusional. This kind of expression helps in gaining the sense of security.

Individuals with deep thoughts and perceptions draw monsters, or whatever they deeply believe in.

Artist with social problems produce more decorative work. Less of a cat and more of carpet. Other artists, with similar problems may not lose their normal work in painting, that being their skill.

Observers of art denote their own inner feelings by projecting their sympathy with the artist, or antipathy, depending on the phobia expressed by the artist and the way that a sentiment has been perceived by an individual.

ART FOR PSYCHOTHERAPY

Whichever the way art is used, it is profound that art may help in organising experience and mastering the environment. Children who paint and model on their pulses help in building their own background. Thinking about it might tie them up. Just let them love art, because art is nature.

The cave man paints the animals of the hunt on the cave walls, to gain imagination, control over them. Man organises nature physically when the wild becomes a garden, or a house replaces the cave. Animals do likewise. Beavers make dams and birds build nests. A wonderful nature.

Sigmund Freud called it the instinct of self-preservation (thanatos). Modern educationists tell us that art is not a luxury, but a necessity. Carl Gustav Jung, the famous Swiss psychotherapist, expressed his belief when he made a study of the primitive people's organising nature and their spear dancing around a hole, which can be accorded to the instict of eros (sex).

How can analytical psychology help us in understanding art? To get a better idea let us look at various artists' paintings from this point of view.

There are two types of mental activity in art. Visual, as expressed by the masters: Velanquez, Courbet and Peplog. The hoptic type of artists includes, Bosch, Klee and Maxwell.

Combined with these, are the masters of art who are categorised according to the Jungian four functions of the psyche.

The thinking type: Ben Nicholson and Barbara Hepworth.

The feeling type: Greco and Renoir.

The sensation type: Van Gogh and Vlaminck.

The intuitive type: Pollock.

These can, also, be divided under the broader terms of introvert and extravert art, again, based on the work of personality types by Jung.

Whatever the motivation for art, be glad of progress, even if it leads away from something. Remember that often enough, the change of 'medicine' will relieve inhibitions. The practising of art, itself, will assist in managing stress, release energy, sublimate the libido and release latent powers.

ART FOR PSYCHOTHERAPY

SECTION THREE: ART FORMS FOR DIAGNOSTIC METHODS USED

Doodles

A doodle is an absent-minded scrawl or scribble, usually executed in some unexpected place, such as the margin of a book or manuscript or a blotting pad when the doodler is preoccupied with some other activity, such as attending a meeting or lecture. The word is supposed to have gained currency because of its use in the film *Mr. Deeds Goes to Town* (1936), though the practice of course is much older, doodles being found in medieval manuscripts, as well as in the notebooks of Leonardo da Vinci and on the margins of manuscripts written by Fyodor Dostoyevsky.

The increasing preoccupation in the 20th century with manifestations of the unconscious and the desire to interpret them both as art forms and as clues to the nature of personality have led to considerable interest in doodles. The Surrealist method of automatic drawing was used by Max Ernst, Salvador Dalí, and André Masson, and Jackson Pollock, an Abstract Expressionist, did a series of drawings that were used as an element in his psychoanalysis.

Sketch

Sketching traditionally is considered to be a rough drawing or painting in which an artist notes down his

ART FOR PSYCHOTHERAPY

preliminary ideas for a work that will eventually be realized with greater precision and detail. The term also applies to brief creative pieces that per se may have artistic merit.

In a traditional sketch, the emphasis usually is laid on the general design and composition of the work and on overall feeling. Such a sketch is often intended for the artist's own guidance; but sometimes, in the context of a bottega (studio-shop) type of production, in which an artist would employ many assistants, sketches were made by the master for works to be completed by others. There are three main types of functional sketches. The first—sometimes known as a croquis—is intended to remind the artist of some scene or event he has seen and wishes to record in a more permanent form. The second—a pochade—is one in which he records, usually in colour, the atmospheric effects and general impressions of a landscape. The third type is related to portraiture and notes the look on a face, the turn of a head, or other physical characteristics of a prospective sitter.

From the 18th century, however, sketch came to take on a new meaning, which has almost come to supersede the traditional one. The emphasis on freshness and spontaneity, which was an integral part of the Romantic attitude, the fact that there was a great increase in the number of amateur artists, and the growing appreciation of nature, accompanied by an expansion of facilities for travel, transformed the sketch into something regarded as an end in itself—a slight and unpretentious picture, in some simple medium (pen and ink, pencil, wash, or watercolour) recording a visual experience. This led to a revaluation of sketches that had originally been created

ART FOR PSYCHOTHERAPY

for other works. Contemporary taste, for instance, tends to value John Constable's sketches as highly as his finished works.

Drawing

A drawing can be the art or technique of producing images on a surface, usually paper, by means of marks, usually of ink, graphite, chalk, charcoal, or crayon.

Drawing as formal artistic creation might be defined as the primarily linear rendition of objects in the visible world, as well as of concepts, thoughts, attitudes, emotions, and fantasies given visual form, of symbols and even of abstract forms. This definition, however, applies to all graphic arts and techniques that are characterized by an emphasis on form or shape rather than mass and colour, as in painting. Drawing as such differs from graphic printing processes in that a direct relationship exists between production and result. Drawing, in short, is the end product of a successive effort applied directly to the carrier. Whereas a drawing may form the basis for reproduction or copying, it is nonetheless unique by its very nature.

Although not every artwork has been preceded by a drawing in the form of a preliminary sketch, drawing is in effect the basis of all visual arts. Often the drawing is absorbed by the completed work or destroyed in the course of completion. Thus, the usefulness of a ground plan drawing of a building that is to be erected decreases as the building goes up. Similarly, points and lines marked on a raw stone block represent auxiliary drawings for the sculpture that will be hewn out of the material. Essentially, every painting is built up of lines and pre-

ART FOR PSYCHOTHERAPY

sketched in its main contours; only as the work proceeds is it consolidated into coloured surfaces. As shown by an increasing number of findings and investigations, drawings form the material basis of mural, panel, and book paintings. Such preliminary sketches may merely indicate the main contours or may predetermine the final execution down to exact details. They may also be mere probing sketches. Long before the appearance of actual small-scale drawing, this procedure was much used for monumental murals. With sinopia—the preliminary sketch found on a layer of its own on the wall underneath the fresco, or painting on freshly spread, moist plaster—one reaches the point at which a work that merely served as technical preparation becomes a formal drawing expressing an artistic intention.

Not until the late 14th century, however, did drawing come into its own—no longer necessarily subordinate, conceptually or materially, to another art form. Autonomous, or independent, drawings, as the name implies, are themselves the ultimate aim of an artistic effort; therefore, they are usually characterized by a pictorial structure and by precise execution down to details.

Formally, drawing offers the widest possible scope for the expression of artistic intentions. Bodies, space, depth, substantiality, and even motion can be made visible through drawing. Furthermore, because of the immediacy of its statement, drawing expresses the draftsman's personality spontaneously in the flow of the line; it is, in fact, the most personal of all artistic statements. It is thus plausible that the esteem in which drawing was held should have developed parallel to the value placed on individual artistic talent. Ever since the

ART FOR PSYCHOTHERAPY

Renaissance, drawing has gradually been losing its anonymous and utilitarian status in the eyes of artists and the public, and its documents have been increasingly valued and collected.

This article deals with the aesthetic characteristics, the mediums of expression, the subject matter, and the history of drawing.

Elements and principles of design

The principal element of drawing is the line. Through practically the entire development of Western drawing, this figure, essentially abstract, not present in nature, and appearing only as a border setting of bodies, colours, or planes, has been the vehicle of a representational more or less illusionist rendition of objects. Only in very recent times has the line been conceived of as an autonomous element of form, independent of an object to be represented.

Conscious and purposeful drawing represents a considerable mental achievement, for the ability to reduce the spatial objects in the world around one to lines drawn on a plane presupposes a great gift for abstraction. The identification of the motif of a drawing by the viewer is no less an achievement, although it is mastered by practically all human beings. The visual interpretation of a line as a representation of a given object is made possible through certain forms of that line that call forth associations. The angular meeting of two lines, for example, may be considered as representing the borders of a plane; the addition of a third line can suggest the idea of a cubic body. Vaulting lines stand for arches, convergent lines for depth.

ART FOR PSYCHOTHERAPY

With the aid of this modest basic vocabulary, one can distil comprehensible images from a variety of linear phenomena. The simple outline sketch—Greek legend has it that the first "picture" originated from copying the shadows on the sand—represents one of the oldest and most popular possibilities of graphic rendition. After decisively characterizing the form of Egyptian drawing and the archaic art of Greece, the outline sketch became the chief vehicle of artistic communication in late antiquity and the Middle Ages. Used in a variety of ways in the early Renaissance, it became dominant once again in Neoclassicism, as it is, for that matter, in the classicist period of a given artist's total work.

The outline sketch is elaborated into the detailed drawing by means of the line, which differentiates between the plastic and the spatial values of the object. Borders of individual objects, changes in the spatial plane, and varying intensities of colour applied within an outline sketch all tend to enrich and clarify the relationship between the whole and its component parts.

The free beginning, the disappearance, or the interruption of a line provides opportunities for gradually slurring an edge until it becomes a plane, for letting colour transitions fade away, for having the line vanish in the depth.

The thickening or thinning of a line can also be used to indicate, spatially or by means of colour, a change in the object designated by that line. Even light-and-shadow values may be rendered by differences in stroke strength.

While the chopping up of a line into several brief segments, and, even more, the drawing of individual lines running parallel in one direction, makes the outlined form

ART FOR PSYCHOTHERAPY

appear less corporeal and firm, it reproduces the visual impact of the form in a more pictorial manner. Slight shifts in the flow of the line are intended to represent smooth curves and transitions; they also reinforce the effect of light striking a surface and thus give the corporeal appearance. Finally, short, curving segments of a line that do not stand in a clearly angular relationship to one another but are arranged on the sheet in loose formation allow the pictorial and colour component to dominate, as in the work of the 16th-century Italian artist Jacopo Tintoretto. An extreme case is the complete dissolution of the linear stroke into dots and spots, as, for example, in the drawings of the 19th-century Pointillist painter Georges Seurat.

A mere combination of these varied shapes of the line, without reference to the mediums in which the lines are drawn, provides the artist with a plethora of subjective opportunities for the expression both of general stylistic traits and of personal characteristics. An arrangement of forceful, mainly straight strokes in accentuated, sharp angles lends the drawing an austere character emphasizing dramatic and expressive traits.

This method of drawing, in fact, is characteristic of stylistic epochs and artistic regions (not to mention individual artists) that prefer these qualities: in the rather sober city of Florence, in German Expressionism, where it is used to convey mood, but also in the drawings of Rembrandt and Vincent van Gogh. Soft lines, on the other hand, running in drawn-out, smoothly rounded forms and stressing graphic regularity above any statement of content, constitute the formal equivalent to elegant, courtly, and lyric qualities of expression.

ART FOR PSYCHOTHERAPY

Accordingly, they are often found in drawings of the Soft style; in the early Renaissance, particularly in the work of artists from the Italian province of Umbria and in young Raphael's sketches; in the work of Nazarenes, a 19th-century group of Romantic painters whose subjects were mainly religious; in the Jugendstil, a late-19th- and early-20th-century German decorative style parallel to Art Nouveau in its organic foliate forms, sinuous lines, and non-geometric curves; and in a very pure form in one of the classic draftsmen, the 19th-century French painter Jean-Auguste-Dominique Ingres.

A markedly even-stroke texture, with waxing and waning strokes in regular proportions and evenly distributed within the page, brings drawing close to calligraphic writing and is found in all stylistic epochs that value ornamentation.

The technique of hatching gives the line an additional potential for the clarification of plastic relationships and of light phenomena. In hatching, parallel, short, equidistant, more or less straight lines create static and tectonic (structural) values by marking individual body planes. Gently curved hatching stresses the roundness of the body and can also accentuate, as tone value, shaded parts of the representation.

Cross-hatching, in which two layers of hatching intersect at right angles, reinforces the body-and-shadow effect. Known since the days of Michelangelo and Dürer in the 15th and 16th centuries, this artistic technique is often used with slanted or even curved hachures for the linear rendition of rounded parts. In rigorously monotone drawings, this method is the most suitable for the depiction of spherical bodies.

ART FOR PSYCHOTHERAPY

The human body, with its highly articulated surface, can be modelled in this fashion very clearly and precisely. For 17th- and 18th-century engravers, this process became the most important means of drawing. All of these different possibilities of linear rendition can be achieved with pen and crayon as well as with the brush.

Plane techniques

Linear techniques of drawing are supplemented by plane methods, which can also be carried out with crayon. For example, evenly applied dotting, which is better done with soft mediums, results in an areal effect in uniform tone. Various values of the chiaroscuro (pictorial representation in terms of light and shade without regard to colour) scale can also be rendered by means of dry or moist rubbing. Pulverized drawing materials that are rubbed into the drawing surface result in evenly toned areas that serve both as a closed foundation for linear drawing and as indication of colour values for individual sections.

More significant for plane phenomena, however, is brushwork, which, to be sure, can adopt all linear drawing methods but the particular strength of which lies in stroke width and tone intensity, a medium that allows for extensive differentiation in colour tone and value. Emphases created by the repeated application of the same tone provide illusionistic indentations that can be conceived of spatially and corporeally. Colour differences result from the use of various mediums. Brushwork also lends itself to spatial and plastic representation, just as it can constitute an autonomous value in non-representational drawings.

ART FOR PSYCHOTHERAPY

All of these effects of monochrome drawing are accentuated with the use of varicoloured mediums of a basic material; for example, coloured chalks, drawing inks, or watercolour. While these mediums enrich the art of drawing, they do not widen its basic range.

Drawing surface

To these graphic elements must be added another phenomenon the formal significance of which is restricted to drawing: the effect of the unmarked drawing surface, usually paper. Almost all studies (drawings of details), many autonomous sheets, most portrait drawings, as well as figure compositions, still lifes, and even landscapes stand free on the sheet instead of being closed off with a frame-line. Thus, the empty surface, suggesting by itself a spatial background to the drawing on it, contributes actively to the artistic effect.

Even within line composition, the surface left blank fulfills an essential role. Among the details conveyed by the empty space may be the planes of a face, the smooth width of a garment, the mass of a figure or object, the substance the borders and nuances of which are indicated by the drawing. Even the space around individual objects, the spatial distance between them and their environment, the width of a river and the depth of a landscape may be merely signalled by the drawing and filled by the void.

This void can itself become the dominant form enclosed by lines or contours—for example, in decorative sketches and in many ornamental drawings that make use of the negative form, an effect attainable also by tinting the blank planes.

ART FOR PSYCHOTHERAPY

Relationship between drawing and other art forms

The bond between drawing and other art forms is of course very close, because the preliminary sketch was for a long time the chief purpose of the drawing. A state of mutual dependence exists in particular between painting and drawing, above all, in the case of sketches and studies for the composition of a picture. The relationship is closest with preliminary sketches of the same size as the original, the so-called cartoons whose contours were pressed through or perforated for dyeing with charcoal dust. Once transferred to the painting surface, the sketch had served its purpose.

On autonomous sheets, too, the close connection between drawing and painting is evidenced by the stylistic features that are common to both. Drawing and painting agree in many details of content and form. Measurements; proportions of figures; relationship of figure to surrounding space; the distribution of the theme within the composition according to static order, symmetry, and equilibrium of the masses or according to dynamic contrasts, eccentric vanishing points, and over-accentuation of individual elements; rhythmic order in separate pictorial units in contrast to continuous flow of lines—all of these formal criteria apply to both art forms.

The uniform stylistic character shared by drawing and painting is often less severely expressed in the former because of the spontaneous flow of the unfettered artist's stroke, or "handwriting", and of the struggle for form as recorded in the pentimenti (indications in the drawing that the artist had changed his mind and drawn over his original formulation). Furthermore drawing can stimulate

ART FOR PSYCHOTHERAPY

certain aspects of movement more easily than painting can through the rhythmic repetition of a contour or the blended rubbing of a sharp borderline.

Still closer, perhaps, is the bond between drawing and engraving, which works with the same artistic means, with monochrome linearity as its main formal element and with various tone and plane methods closely related to those of drawing.

Drawing is more independent than sculpture because sculpture uses a three-dimensional model. As a result, sculptors' drawings can always claim a greater degree of autonomy.

Surfaces

One can draw on practically anything that has a plane surface (it does not have to be level)—for example, papyrus and parchment, cloth, wood, metals, ceramics, stone, and even walls, glass, and sand. (With some of these, to be sure, another dimension is introduced through indentations that give the visual effect of lines.) Ever since the 15th century, however, paper has been by far the most popular ground.

The technique of paper manufacturing, introduced from East Asia by the Arabs, has remained virtually unchanged for the past 2,000 years. A fibrous pulp of mulberry bark, hemp, bast, and linen rags is drained, pressed, and dried in flat moulds. The introduction of wood pulp in the mid-19th century, which enabled manufacturers to satisfy the enormously increased demand for bulk paper, did not affect art paper because paper of large wood content yellows quickly and is therefore ill-suited for art drawing.

ART FOR PSYCHOTHERAPY

The essential preparation of the paper to give it a smooth and even surface for writing or drawing was once done by rubbing it with bone meal, gypsum chalk, or zinc and titanium white in a very thin solution of glue and gum arabic. The proper priming, achieved through repeated rubbing and polishing, was of the utmost importance, especially for metalpoint drawings. If such preparation is too weak, the paper accepts the stroke badly; if it is too strong, the coating cracks and chips under the pressure of the hand.

Since the early 15th century, however, the sheets have been given the desired smooth and non-absorbent consistency by dipping them in a glue or alum bath.

The addition of glue also made it possible to impart to the pulp paper a quality that permitted pen drawings. Pigments, too, could of course be added to the pulp, and the so-called natural papers—chiefly blue and called Venetian papers after the centre of the retail trade in this commodity—became more and more popular. While the 17th century liked half tints of blue, grey, brown, and green, the 18th preferred warm colours such as ivory and beige, along with blue. Since the 18th century, paper has been manufactured in all conceivable colours and half tones.

The range of quality has also greatly increased since the end of the 18th century to give more painstakingly produced drawing papers. Even in earlier times, the absorbent Japan paper made of mulberry bark enjoyed great popularity. Handmade paper, stronger and free of wood, with an irregular edge, has remained to this day a favourite surface for drawings.

ART FOR PSYCHOTHERAPY

Vellum, delicate and without veins, resembles parchment in its smooth surface. Modern watercolour paper is a pure linen paper glued in bulk and absolutely free of fat and alum; its two surfaces are of different grain. For pastel drawings, a firm, slightly rough surface is indicated, whereas pen drawings are best done on a very smooth paper.

Granulated and softer drawing tools, such as charcoal, chalk, and graphite are not as dependent on a particular type of paper; but, because of their slight adhesiveness, they often require a stronger bond with the foundation as well as some form of surface protection. This process of fixing was formerly done through repeated varnishing with gum-arabic solution and even with glue or egg-white emulsion. Modern siccatives (drying substances) inhibit discoloration but cannot prevent the living surface from appearing sealed, as it were, under a skin.

In pastels especially, the manifold prismatic effects of finely powdered coloured crayons are thus lost, and the bright and airy surface is turned into an amorphous, heavy layer. Pastels, which brush off easily, are therefore best preserved under glass.

Tools and techniques

Such varied tools as slate pencils, charcoal, metal styli, and chalks may be used for drawing as well as all writing utensils, including pens, pencils, and brushes. Indeed, even chisels and diamonds are used for drawing, not to mention saws, drills, and fire.

Dry drawing tools differ in effectiveness from liquid ones because it is not irrelevant from the artistic point of view

ART FOR PSYCHOTHERAPY

whether one uses a self-drawing medium that permits an evenly flowing line dependent only on hand pressure or a transferring tool that must be put down periodically and refilled, with resultant differences in the strength and concentration of the line. Modern drawing mediums that combine both possibilities, such as fountain pens, ball-point pens, and fibre-tipped pens, are recent inventions.

No less varied than the nature and composition of these drawing mediums is their aesthetic effect. It would nevertheless be wrong to systematize the art of drawing on the basis of the techniques applied; not only does almost every technique have several applications but it can also be combined with other techniques, and the draftsman's temperament inevitably plays a role as well. Even if certain techniques predominate in certain periods, the selection of drawing mediums depends on the intended effect and not vice versa.

Artists have always been able to attain the desired effect with a variety of techniques. Dry mediums, for example, are predestined for clear lines, liquid ones for plane application. Yet extremely fine strokes can also be made by brush, and broad fields can be marked in with pencil or crayon. Some mediums, including charcoal, one of the oldest, if not the oldest of all, allow both extremes.

Charcoal

In every hearth or fireplace, partially consumed pieces of wood remain that can be used as a convenient tool for drawing. Evidence of charcoal sketches for mural, panel, and even miniature paintings can still occasionally be seen under the pigment.

ART FOR PSYCHOTHERAPY

Drawing charcoal produced from wood that is as homogeneous as possible gives a porous and not very adhesive stroke. The pointed charcoal pencil permits hair-thin lines; if used broadside on the surface, it creates evenly toned planes.

Rubbing and pulverizing the charcoal line results in dimmed intermediate shades and delicate transitions. Because of its slight adhesiveness, charcoal is eminently suited to corrective sketching; but if the drawing is to be preserved, it must be protected by a fixative.

As a medium for quick, probing sketches and practice in studying models, charcoal was once much used in all academies and workshops. The rapid notation of difficult poses, such as Tintoretto demanded of his models, could be done quickly and easily with the adaptable charcoal pencil. While some of these sheets were deemed worthy of preservation, hundreds have surely been lost.

Charcoal has often been used for portrait drawings to preserve for the eventual painting pictorial tints that were already present in the preliminary sketch. When destined to be autonomous portraits, charcoal drawings are executed in detail; with their sharp accents and delicate modelling, such portraits cover the whole range of the medium. In "Portrait of a Lady", by the 19th-century French painter Édouard Manet, the grain of the wood in the chair, the fur trimming on the dress, the compactness of the coiffure, and the softness of the flesh are all rendered in the same material: charcoal.

Popular as that material was for studies and sketches, it has been used for independent drawings destined for preservation by only a few artists; for example, the 17th-century Dutch painter Paulus Potter. It is somewhat more frequent among the great draftsmen of the 19th and 20th centuries, such as Edgar Degas, Henri de Toulouse-Lautrec, Käthe Kollwitz, and Ernst Barlach.

Oiled charcoal, with the charcoal pencils dipped in linseed oil, provides better adhesion and a deeper black. Used in the 16th century by Tintoretto, this technique was applied above all by the Dutch draftsmen of the 17th century in order to set deep-black accents.

The advantage of better adhesion in the indentations of the paper in contrast to dry charcoal, which sticks to the elevations, has to be paid for, however, by "incorrigibility"; *i.e.*, correction cannot be made. In addition, charcoal crayons that have been deeply dipped in oil show a brownish streak left by the oil alongside the lines; this technique was used in the 20th century by the American artist Susan Rothenberg.

Chalks

The chalks, which resemble charcoal pencils in outward appearance, are an equally important drawing medium. If charcoal was primarily a medium for quick sketching that could be corrected and for the search for artistic form, chalk drawing, which can also fulfil all of these functions, has steadily gained in importance as an autonomous vehicle of expression.

Since the end of the 15th century, stone chalk, as found in nature, has become increasingly more significant in art drawing. As a basic material, alumina chalk has various degrees of hardness, so that the stroke varies from slightly granular to homogeneously dense and smooth.

ART FOR PSYCHOTHERAPY

The attempt to produce a crayon or pencil of the greatest possible uniformity has led to the production of special chalks for drawing; that is, chalks, which, after being pulverized, washed, and moulded into convenient sticks, allow a softer and more regular stroke and are also free of sandy particles.

The admixture of pigments (carbons in the case of black chalks) creates various tints from a rich black to a brownish grey; compared to the much-used black chalk, the brown variety is of little significance. White chalk, also found in nature, is rarely employed as an independent medium for drawing, although it is frequently used in combination with other mediums in order to achieve reflections of light as individual accents of plastic modelling.

Beginning with the 15th century, chalk has been used increasingly for studies and sketches. Its suitability for drawing exact lines of any given width and also for laying on finely shaded tints makes it particularly appropriate for modelling studies.

Accents that stress plastic phenomena are applied by varying the pressure of the hand. Characteristic details in portrait drawings in particular can be brought out in this manner. Pictorial values as well as light and shadow effects can be rendered with chalk without losing their firm, plastic form. For the same reason, chalk is also most valuable in sketching out paintings and indicating their values.

All of these qualities explain why chalk is such a good medium for autonomous drawings. Indeed, there is scarcely a draftsman who has not worked in chalk, often in combination with other mediums. Aside from portrait

ART FOR PSYCHOTHERAPY

drawings done all over the world, landscapes have formed the main theme of chalk drawings, especially with the Dutch, in whose art landscape drawings have played a large role.

Ever since the invention of artificial chalk made of lampblack (a fine, bulky, dull-black soot deposited in incomplete combustion of carbonaceous materials), which possesses a precisely measurable consistency—an invention ascribed to Leonardo da Vinci—the pictorial qualities of chalk drawing have been fully utilized. Chalks range from those that are dry and charcoal-like to the fatty ones used by lithographers.

Another very important drawing pencil is similarly a chalk product: the red pencil, or sanguine, which contains ferric oxide, which occurs in nature in shadings from dark brown to strong red and can also be manufactured from the same aluminum-oxide base with ferric oxide or rust added. Besides the stronger pictorial effect possible because of its chromatic value, sanguine also possesses a greater suppleness and solubility in water. Thus, a homogeneous plane can be created through moist rubbing, a compact stroke through liquid linear application, a very delicate tone through light wiping.

Although this oxide was used for red tints in prehistoric painting, sanguine does not seem to have acquired artistic dignity until the 15th century, when it became customary to fix drawings by painting them over with a gum solution, for sanguine has no more adhesiveness than charcoal. In the 15th century, sanguine was a popular drawing medium because of its wealth of pictorial possibilities.

ART FOR PSYCHOTHERAPY

Those inclined to be colorists—such as the portraitists Jean Clouet and Hans Holbein, the Flemish painters around Peter Paul Rubens, and, above all, the French artists of the 18th century—particularly favoured it. The possibilities of sanguine range from suggestive forms with markedly plastic values to a very pictorial, soft rendition of visual surface stimuli.

A combination of various chalks offers still richer colouristic possibilities. Black chalk and sanguine have been widely used since the 16th century to achieve colour differentiation between flesh tones, hair, and the material of garments. The combination of black and white chalk serves plastic modelling, as does that of the softer sanguine with white chalk; in the former case, the accentuation rests with the black, in the latter, with the more suggestive delineation in white.

A decidedly colouristic method lies in the combination of various chalk colours with one another and with tinted paper. Such pictorially executed sheets, called *à deux crayons* (with two colours) and *à trois crayons* (with three colours), respectively, were especially popular in 17th- and 18th-century France. Antoine Watteau reached a previously unheard of harmony of different chalks on natural paper. With the three colours, Nicolas Lancret, Jean-Étienne Liotard, Jacques-André Portail, François Boucher—to name but a few such artists—achieved sensitive drawings that are very appealing coloristically.

An additional colour refinement is made possible with pastel crayons. An ample selection of dry colour pigments in pastel crayons, prepared with a minimum of agglutinants and compounded with different shades of

ART FOR PSYCHOTHERAPY

white for the articulation of tints, is commercially available.

The colours can be laid on in linear technique directly with the crayons, but an area application made with a piece of soft suede or directly, with the fingers, is more frequent. Although this technique was known to the Accademia degli Incamminati (to the painter Guido Reni, for example) as early as the 17th century, it did not reach its flowering until the 18th century, especially in France (with Jean-Marc Nattier and Jean-Baptiste-Siméon Chardin) and in Venice (with Rosalba Carriera). Pastel chalks are particularly favoured for portraits; their effect approximates that of colour-and-area painting rather than line drawing.

In the 19th and 20th centuries, Degas reverted to a stronger accentuation of the delineatory aspects of drawing. With intermediate varnishes he achieved an overlay drawing with different colours and thus an increased emphasis on individual strokes. This technique, fundamentally different from the older one, was imitated with minor variations by Odilon Redon, Gustave Moreau, Jean-Édouard Vuillard, Pierre Bonnard, and others. It has also been borrowed by such Expressionist artists as Edvard Munch and Ernst Ludwig Kirchner.

Modern grease chalks offer a chromatic scale of similar range. Developed originally for such technical purposes as the lettering of very smooth surfaces, such as metal or glass, they can be applied in the same flat manner as pastels, although with the opposite aesthetic effect: that of compact colours. It was the 20th-century English sculptor Henry Moore who first and convincingly

ART FOR PSYCHOTHERAPY

exploited the feasibility of continuing, with other mediums, such as pen or watercolour, work on the firm surface that had been led out with grease chalks.

Metalpoints

Metalpoints have been used for writing and delineation ever since the scriptoria of antiquity. It required little imagination to employ them also in drawing. The most frequently used material was soft lead, which on a smooth surface comes out pale grey, not very strong in colour, and easily erasable but very suitable for preliminary sketches. Aside from lead, tin and copper were also used, as well as sundry lead-and-pewter alloys.

The 15th-century Venetian painter Jacopo Bellini's book of sketches in London with lead point drawings on tinted paper is a particularly valuable example of this technique, even if individual portions and, indeed, entire pages that had become effected were drawn over long ago. One can see little more than the traces left by the pencil because, as in many other metalpoint drawings, the sketches were redrawn in another medium. Botticelli, for example, sketched with a leadpoint the outline of his illustrations to Dante's *Divine Comedy,* retracing them afterward with the pen. Metalpoints were used into the 18th century for perspectivist constructions and auxiliary delineation, especially in architectural drawings.

More suited to permanent drawing is the silverpoint, which requires special preparation of the foundation and, once applied, cannot be corrected. Its stroke, also pale gray, oxidizes into brown and adheres inerasably.

ART FOR PSYCHOTHERAPY

Silverpoint drawings accordingly require a clearer concept of form and a steady hand because corrections remain visible. Because too much pressure can bring about cracks in the foundation, the strokes must be even; emphases, modelling, and light phenomena must be rendered either by means of dense hachures, repetitions, and blanks or else supplemented by other mediums. Despite these difficulties, silverpoint was much used in the 15th and 16th centuries.

Dürer's notebook on a journey to Holland shows landscapes, portraits, and various objects that interested him drawn in this demanding technique. Silverpoint was much in demand for portrait drawings from the 15th into the 17th century; revived in the 18th-century Romantic era, it was also used by modern artists, most notably Picasso.

Graphite point

Toward the end of the 16th century, a new drawing medium was introduced and soon completely displaced metalpoint in sketching and preliminary drawing: the graphite point. Also called Spanish lead after its chief place of origin, this drawing medium was quickly and widely adopted; but because of its soft and smeary consistency it was used for autonomous drawings only by some Dutch painters, and even they employed it mostly in conjunction with other points.

(It might be added that the graphite point was originally taken for a metal because its texture shines metallically in slanting light.)

ART FOR PSYCHOTHERAPY

The lead pencil, or more properly *crayon Conté*, became established in art drawing after Nicolas-Jacques Conté invented, around 1790, a manufacturing process similar to that used in the production of artificial chalk. Purified and washed, graphite could henceforth be made with varying admixtures of clay and in any desired degree of hardness.

he hard points, with their durable, clear, and thin stroke layers, were especially suited to the purposes of Neo-classicist and Romantic draftsmen. The Germans working in Rome, in particular, took advantage of the chance to sketch rapidly and to reproduce, in one and the same medium, subtle differentiations as well as clear proportions of plasticity and light.

Among the most masterful pencil artists of all was Ingres, who pre-sketched systematically in pencil the well-thought-out structure of his paintings.

The more pictorially inclined artists of the late 19th century preferred softer pencils in order to throw into plastic relief certain areas within the drawing. Seurat, on the other hand, reached back to graphite in his drawings from the concert cafés, among them *At the Concert Européen*, in which he translated the Pointillistic technique (applying dots of colour to a surface so that from a distance they blend together) into the monochrome element of drawing.

Pencil frottage (rubbing made on paper laid over a rough surface), first executed by the Surrealist artist Max Ernst, represents a marginal kind of drawing, for here the artist's hand is no longer the sole creator of forms.

ART FOR PSYCHOTHERAPY

Coloured crayons

Coloured crayons, in circulation since the late 19th century, offer all the possibilities of black graphite points; and, in combinations, they attain a stronger colour value than chalks because they do not merge with one another. Every line preserves its original and characteristic colour, a form of independence that Gustav Klimt and Picasso exploited to the full.

Incised drawing

A role apart is that played by incised drawings. Their pronounced linearity gives them the visual appearance of other drawings; materially, however, they represent the opposite principle, that of subtracting from a surface rather than adding to it. Incised drawings are among the oldest documents of human activity. In primitive African cultures, the methods and forms of prehistoric bone and rock drawings have survived into the present.

In a decorative and conceivably also symbolic form, incised decorations on pottery have existed for thousands of years; insofar as the comparison is valid, they correspond in every formal respect to applied drawings of the same period.

A formal equivalent may also be observed in later times: in the decorative details of implements, especially metal—from the drawings on Greek mirrors, through the jewelry made at the end of the Roman Empire, to the scenes on medieval weapons and, above all, on Renaissance dress armour.

ART FOR PSYCHOTHERAPY

More often than not these are drawings that follow certain models rather than free drawings in the sense of sketches.

Logically, one would also have to consider all niello work under the heading of drawing, because the picture in this case is cut out of the metal and filled with a deep black-coloured paste so that it appears to the eye as a linear projection on a plane. In like manner, work with the graver or burin (cutting tools) and with the etching needle on the engraving plate may be considered to parallel in its execution that gradual effort applied directly to the carrier that was defined earlier as the art of drawing.

The difference lies in the fact that this work is not a goal in itself but the prerequisite for a printing process that is intended to be repetitive.

Brush, pen, and dyestuffs

Of the many possibilities of transferring liquid dyestuffs onto a plane, two have become particularly significant for art drawing: brush and pen. To be sure, finger painting, as found in prehistoric cave paintings, has occasionally been practiced since the late Renaissance and increasingly so in more recent times.

For drawing as such, however, the method is irrelevant. Similarly, the use of pieces of fur, frayed pieces of wood, bundles of straw, and the like is more significant as a first step toward the camel's-hair brush than as indication that these objects were ever drawing mediums in their own right.

Although it is antedated by the brush, which in some cultures (East Asia, for example) has remained in

ART FOR PSYCHOTHERAPY

continued use, the pen has been the favorite writing and drawing tool ever since classical antiquity.

Pens

The principle of transferring dyestuffs with the pen has remained virtually unchanged for thousands of years. The capillary effect of the split tip, cut at a slant, applies the drawing fluid to the surface (parchment, papyrus, and, since the late Middle Ages, almost exclusively paper) in amounts varying with the saturation of the pen and the pressure exerted by the drawing hand.

The oldest form is that of the reed pen; cut from papyrus plants, sedge, or bamboo, it stores a reservoir of fluid in its hollow interior. Its stroke—characteristically powerful, hard, and occasionally forked as a result of stronger pressure being applied to the split tip—became a popular medium of artistic expression only with the rise of a subjective view of the artist's personality during the Renaissance.

Rembrandt made superb use of the strong, plastic accents of the reed pen, supplementing it as a rule with other pens or brushes. Beginning in the 19th century with the Dutch artist van Gogh, pure reed-pen drawings with a certain forcefulness of expression have been created by many artists. Expressionists such as George Grosz used the reed pen frequently.

If the selection of the reed pen already implies a formal statement of sorts, which of the quill pen opens up a far wider range of possibilities. Ever since the rise of drawing in Western art—that is, since the late Middle Ages—the

ART FOR PSYCHOTHERAPY

quill has been the most frequently used instrument for applying liquid mediums to the drawing surface.

The importance accorded to this tool is attested by the detailed instructions in painters' manuals about the fashioning of the pen from wing shafts of geese, swans, and even ravens. The supple tip of the quill, available in varying strengths, permits a relatively wide scale of individual strokes—from soft, thin lines, such as those used in preliminary sketches for illustrations in illuminated books, through waxing and waning lines that allow differentiation within the stroke, to energetic, broad lines. It was only when metal pens began to be made of high-grade steel and in different strengths that they became a drawing implement able to satisfy the demands made by the individual artist's hand.

Inks

Although all dyestuffs of low viscosity lend themselves to pen drawing, the various inks are most often employed. The manufacture of gallnut ink had been known from the medieval scriptoria (copying rooms set apart for scribes in monasteries). An extract of gallnuts mixed with iron vitriol and thickened with gum-arabic solution produces a writing fluid that comes from the pen black, with a strong hint of purple violet, and dries almost black. In the course of time it turns a darkish brown, so that the writing fluid in old manuscripts and drawings cannot always be identified by the colour alone.

In contrast to other brown writing fluids, the more strongly coloured parts of gallnut ink remain markedly darker; and because inks of especially great vitriol content decompose the paper, the drawing, particularly in its more coloured portions, tends to shine through on the reverse side of the sheet. Only industrially produced chemical inks possess the necessary ion balance to forestall this undesirable effect.

Another ink, one that seems to have found no favour as a writing fluid but has nonetheless had a certain popularity in drawing, is bistre, an easily dissolved, light-to-dark-brown transparent pigment obtained from the soot of the lampblack that coats wood-burning chimneys. Its shade depends both on the concentration and on the kind of wood from which it is derived, hardwoods (especially oaks) producing a darker shade than conifers, such as pine.

During the pictorially oriented Baroque period, in the 17th and early 18th centuries, the warm tone that can be thinned at will made bistre a popular medium with which to supplement the planes of a pen drawing.

Also derived from a carbon base is India ink, made from the soot of exceptionally hard woods, such as olive or grape vines, or from the fatty lampblack of the oil flame, with gum-arabic mixed in as a binding agent. This deep-black, thick fluid preserves its dark tone for a long time and can be thinned with water until it becomes a light grey. Pressed into sticks or bars, it was sold under the name of Chinese ink or India ink.

This writing fluid, known already in Egypt and used to this day in China and India, has been manufactured in Europe since the 15th century. Favoured in particular by German and Dutch draftsmen because of its strong colour, it lent itself above all to drawing on tinted paper. Since the 19th century, India ink has been the most popular drawing ink for pen drawings, replacing all other

ART FOR PSYCHOTHERAPY

dyestuffs in technical sketches. Only very recently have writing inks gained some significance in art drawing—in connection with the practical fountain pen.

For a relatively short time, a dyestuff of animal origin, sepia, obtained from the pigment of the cuttlefish, was used for drawing. Known since Roman times, it did not come into general use until the 18th century. Compared to yellowish bistre, it has a cooler and darker tone, and is brown with a trace of violet. Until the 18th century, it was employed by such amateur painters as the poet Goethe because of its effectiveness in depth; as a primary pigment, however, it has been completely replaced by industrially manufactured watercolours.

Other dyestuffs are of only minor importance compared with these inks, which are primarily used for pen drawings. Minium (red lead) was used in the medieval scriptoria for the decoration of initial letters and also in illustrated pen drawings.

Chinese white is easier to apply with a pointed brush because of its thickness; other pigments, among them indigo and green copper sulphate, are rarely found in drawings. For them, too, the brush is a better tool than the pen. The systematically produced watercolours of various shades are almost wholly restricted to technical drawings.

Pen drawings

In combination with written texts, pen drawings are among the oldest artistic documents. Already in classical times, texts were illustrated with firm contours and sparse interior details. During the Middle Ages, marginal

ART FOR PSYCHOTHERAPY

drawings and book illustrations were time and again pre-sketched, if not definitively executed, with the pen. In book painting, decidedly delineatory styles developed in which the brush was also employed in the manner of a pen drawing: for example, in the Carolingian school of Reims, which produced the Utrecht Psalter in the 9th century, and also in southern Germany, where a separate illustrative form with line drawings was widespread with the *Biblia Pauperum* ("Poor People's Bibles," biblical picture books used to instruct large numbers of people in the Christian faith).

The thin-lined outline sketch is also characteristic of the earliest individual drawings of the late Middle Ages and early Renaissance. Sketches after ancient sculptures or after nature as well as compositions dealing with familiar motifs form the main themes of these drawings. Such sheets were primarily used as models for paintings; gathered in sketchbooks, they were often handed on from one generation to the next.

The practical usefulness of these drawings is attested by the supplements added to them by younger artists and by the fact that many metalpoint drawings that had become hard to decipher were redrawn with the pen, as shown by the sketchbooks of the 15th-century Italian artist Antonio Pisanello, now broken up and preserved in several different collections.

In the 16th century, the artistic range of the pen drawing reached an individual articulation that it hardly ever attained again. Every artist was free to exploit with the pen the formal possibilities that corresponded to his talents. Thus Leonardo used a precise stroke for his scientific drawings; Raphael produced relaxed sketches,

ART FOR PSYCHOTHERAPY

in which he probed for forms and variations of form; Michelangelo drew with short strokes reminiscent of chisel work; Titian contrasted light and dark by means of hachures laid broadly over the completed figures.

Among the Northerners, Dürer mastered all the possibilities of pen drawing, from quick notation to the painstakingly executed autonomous drawing, ranging from a purely graphic and delineatory technique to a spatial and plastic modelling one; it is no wonder that he stimulated so many other artists. The subjective attitude of the later 16th century is often expressed more clearly in Mannerist drawings—characterized by spatial incongruity and excessive elongation of the human figures, which are as revelatory of the artist's personality as handwriting—than it is in completed works of painting and sculpture.

A special form of exact drawing is found in models for engravings; some of these were directly mounted on the wood block; some anticipate the style of the copperplate engraving in the pen-drawing stage, with waxing and waning lines, delicate stroke layers, and cross-hatching for spatial and plastic effects.

In the 17th century, the pen drawing took second place to combined techniques, especially wash, a sweep or splash of colour, applied with the brush. An open style of drawing that merely hints at contours, along with contrasting thin and powerful strokes, endowed the line itself with expressive qualities. In his numerous drawings, Rembrandt in particular achieved an exceedingly subtle plastic characterization and even light values through the differentiation of stroke layers and the combination of various pens and brushes.

ART FOR PSYCHOTHERAPY

Additional techniques came to the fore in the 18th century, with the pen sketch providing the scaffold for the drawing that was carried out in a pictorial style. Only decorative sketches and practical studies were laid out more often as linear drawings.

The closed, thin-contour drawing regained its importance with Neoclassicism at the end of the 18th century. The Nazarenes (the nickname of the Lucas Brotherhood—later Guild of St. Luke, who lived in monastic style) and Romantics consciously referred to the early Renaissance manner of drawing, modelling with thin lines. With closed contours, carefully set hair-and-shadow strokes, and precise parallel hachures, they attained plastic values by purely graphic means.

This technique was again followed by a more pictorially oriented phase, culminating in the late 19th century in the recognition of drawing as the most immediate and personal expression of the artist's hand. The pure pen drawing took its place by the side of other highly esteemed art forms.

The English Art Nouveau artist Aubrey Beardsley at the end of the 19th century applied the direct black–white contrast to planes, while in the 20th century the French masters Henri Matisse and Picasso reduced the object to a mere line that makes no claim to corporeal illusion.

A large number of illustrators, as well as the artists who draw the comic strips, prefer the clear pen stroke. In the Russian artist Wassily Kandinsky's non-representational compositions, finally, the independence of the line as an autonomous formal value became a new theme in drawing.

ART FOR PSYCHOTHERAPY

In the hair-thin automatist seismograms (so-called because of their resemblance to the records of earthquakes) of the 20th-century German artist Wols (Alfred Otto Wolfgang Schulze), which are sensitive to the slightest stirring of the hand, this theme leads to a new dimension transcending all traditional concepts of a representational art of drawing.

Brush drawings

Although the brush is best suited to the flat application of pigments—in other words, to painting—its use in a clearly delineatory function, with the line dominating and (a crucial property of brush drawing) in monochrome fashion, can be traced back to prehistoric times.

All of the above-mentioned drawing inks have been used as dyes in brush drawings, often with one and the same pigment employed in combined pen-and-brush work. Still greater differentiation in tone is often obtained through concentrated or thinned mediums and with the addition of supplementary ones.

To the latter belong chiefly distemper, a paint in which the pigments are mixed with an emulsion of egg or size or both, and watercolours, which can be used along with bistre and drawing ink. Even oils can sometimes be used for individual effects in drawing, as in the works of Jacob Jordaens.

Sinopia, the preliminary sketch for a monumental wall painting, was done with the brush and has all the characteristics of a preparatory, form-probing drawing. The sketch was carried out directly on the appropriate

ART FOR PSYCHOTHERAPY

spot and covered over with a thin layer of plaster, on which the pictorial representation was then painted.

The brush drawing differs from the pen drawing by its greater variation in stroke width, and by the stroke itself, which sets in more smoothly and is altogether less severely bordered. Early brush drawings nonetheless show a striking connection with the technique of the pen drawing. The early examples of the 15th century completely follow the flow of contemporaneous pen drawings.

Leonardo's or Dürer's pen drawings, with their short, waxing and waning stroke layers, refine the system of pen drawing; many 16th-century artists used a comparable technique. The brush drawing for chiaroscuro sheets on tinted paper was popular because Chinese white, the main vehicle of delineation in this method, is more easily applied with the brush than the pen and because the intended pictorial effect is more easily attained, thanks to the possibility of changing abruptly to a plane representation.

Such representations are particularly distinctive as done by Vittore Carpaccio and Palma il Giovane in Venice and in a Mannerist spotting technique used by Parmigianino. In the 16th century, the brush nevertheless played a greater role as a supporting than as an independently form-giving instrument. Pure brush drawings were rare even in the 17th century, although the brush played a major role in landscapes, in which, by tinting of varying intensity, it ideally fulfilled the need to provide for all desired degrees of spatial depth and strength of lighting.

Dutch artists, such as Adriaen Brouwer, Adriaen van Ostade, and Jan Steen, as well as the French artist

ART FOR PSYCHOTHERAPY

Claude Lorrain, transcended the limits of drawing in the narrower meaning of the term by doing brushwork limited to a few tones within a monochrome scale, giving the impression of a pictorial watercolour.

Although the coloristically inclined 18th century was little interested in the restriction to a few shadings within one colour value, Jean-Honoré Fragonard raised this technique to perfection, with all its possibilities of sharply accented contours, soft delineation, delicate tones, and deep shadows.

The brush drawings of the Spanish painter Francisco Goya must also be counted among the great achievements of this technique. In his strong plastic effects, the English painter George Romney made the most of the contrast between the white foundation and the broad brushstrokes tinted in varying intensities.

Other English artists, among them Alexander Cozens, John Constable, and J.M.W. Turner, took advantage of the delicately graded pictorial possibilities for their landscape studies.

In the 19th century, the French artists Théodore Géricault, Eugène Delacroix, and Constantin Guys still followed the character of the brush drawing, even though it was already being replaced by the variegated watercolour and gouache painting, a method of painting with opaque colours that have been ground in water and mingled with a preparation of gum.

In modern drawing, the brush has regained some importance as an effective medium for contrasting planes and as carrier of the theme; in this, the dry brush has proven itself a useful tool for the creation of a granular surface structure.

ART FOR PSYCHOTHERAPY

Combinations of various techniques

The combination of various techniques plays a greater role in drawing than in all other art forms. Yet it is necessary, in the numerous drawings in which two or more mediums are involved, to distinguish between those in which the mediums were changed in the course of artistic genesis and those in which an artistic effect based on a combination of mediums was intended from the beginning.

In the first case, one is confronted with a preliminary sketch, as it were, of the eventual drawing: the basic structure with some variations is tried out in charcoal, chalk, metalpoint, pencil, or some other (preferably dry and easily corrected) material and then carried out in a stronger and more durable medium. Most pen drawings are thus superimposed on a preliminary sketch. The different materials actually represent two separate stages of the same artistic process.

More relevant artistically is the planned combination of different techniques that are meant to complement each other. The most significant combination from the stylistic point of view is that of pen and brush, with the pen delineating the contours that denote the object and the brush providing spatial and plastic as well as pictorial—that is, colour—values.

The simplest combined form is manuscript illumination, where the delineated close contours are filled in with colour. The drawing may actually be improved if this is

ART FOR PSYCHOTHERAPY

done by a hand other than the draftsman's or at a later time.

More important is brushwork that supplements linear drawing, in which entire segments may be given over to one technique or the other; for example, the considerable use of white (which is hard to apply with the pen) in drawings on tinted paper. In similar complementary fashion the brush may be used for plastic modelling as a way of highlighting, that is indicating the spots that receive the greatest illumination.

The technique of combined pen-and-brush drawing was favoured by the draftsmen of Germany and the Netherlands, especially in the circle around Dürer and the south German Danube School. Shadows, too, can be inserted in a drawing with dark paint. The illusion of depth can also be achieved with white and dark colours in a pure chalk technique.

In contrast to these methods, which still belong to a linear system of drawing, is the flat differentiation of individual segments of a work in (usually) the same medium: wash. Various bodies and objects are evenly tinted with the brush within or along the drawn contours.

Planes are thus contrasted with lines, enhancing the illusionary effect of plasticity, space, and light and shadow. This modelling wash has been used again and again since the 16th century, sometimes in combination with charcoal, chalk, or pencil drawings.

A further refinement, used particularly in landscape drawings, is wash in varying intensities; additional shadings in the sense of atmospheric phenomena, such as striking light and haze merging into fog and cloud, can

ART FOR PSYCHOTHERAPY

be rendered through thinning of the colour or repeated covering over a particular spot.

A chromatic element entered drawing with the introduction of diluted indigo, known in the Netherlands from the East India trade; it is not tied to objects but used in spatial and illusionist fashion, by Paul Brill and Hans Bol in the 16th and 17th centuries, for example.

The mutual supplementation and correlation of pen and brush in the wash technique was developed most broadly and consistently in the 17th century, in which the scaffold, so to speak, of the pen drawing became lighter and more open, and brushwork integrated corporeal and spatial zones.

The transition from one technique to the other—from wash pen drawings to brush drawings with pen accents—took place without a break. Claude Lorrain and Nicolas Poussin in 16th- and 17th-century France are major representatives of the latter technique, and Rembrandt once again utilized all its possibilities to the full.

Whereas this method served—within the general stylistic intentions of the 17th century—primarily to elucidate spatial and corporeal proportions, the artists of the 18th century employed it to probe this situation visually with the aid of light. The unmarked area, the spot left empty, has as much representational meaning as the pen contours, the lighter or darker brush accent, and the tinted area.

The art of omission plays a still greater role, if possible, in the later 19th century and in the 20th. Paul Cézanne's late sheets, with their sparse use of the pencil and the carefully measured out colour nuances, may be considered the epitome of this technique. As the

ART FOR PSYCHOTHERAPY

colouring becomes increasingly varied through the use of watercolours to supplement a pen or metalpoint drawing, one leaves the concept of drawing in the strict sense of the term.

According to the quality and quantity of the mediums employed, one then speaks of "drawings with watercolour," "water-colourized drawings," and "watercolours on preliminary drawings." The predominant stroke character, rather than the fact that paper is the carrier, is the chief feature when deciding whether or not the work may legitimately be called a drawing.

The combination of dry and fluid drawing mediums provides a genuine surface contrast that may be exploited for sensuous differentiation. Here again a distinction must be made between various ways of applying the identical medium—for example, charcoal and charcoal dust in a water solution or, more frequently, sanguine and sanguine rubbed in with a wet brush—and the stronger contrast brought about by the use of altogether different mediums.

Chalk drawings are frequently washed with bistre or watercolour, after the principle of the washed pen drawing. Stronger contrasts, however, can be obtained if the differing techniques are employed graphically, as the Flemish draftsmen of the 17th century liked to do. The Chinese ink wash of chalk drawings also contributes to the illusion of spatial depth.

Along with such Dutch painters as Jan van Goyen and members of the family van de Velde, Claude Lorrain achieved great mastery in this technique. The differentiated treatment of the foreground with pen and

ART FOR PSYCHOTHERAPY

brush and the background with chalk renders spatial depth plausible and plastic.

In modern art, the use of different mediums—whether for plastic differentiation, such as Henry Moore carried out with unequalled mastery in his "Shelter Drawings," or only for the purpose of contrasting varied surface stimuli of non-representational compositions as well as the enrichment with colours and even with collage elements (the addition of paper, metal, or other actual objects) broadens the concept of the drawing so that it becomes an autonomous picture the mixed technique of which transcends the borderline between drawing and painting.

Mechanical devices

Mechanical aids are far less important for art drawing than for any other art form. Many draftsmen reject them altogether as un-artistic and inimical to the creative aspect of drawing.

Apart from the crucial importance that mechanical aids have had and continue to have for all kinds of construction diagrams, plans, and other applied drawings, some mechanical aids have been used in varying but significant measure for artistic drawings.

The ruler, triangle, and compass as basic geometric instruments have played a major role, especially in periods in which artists created in a consciously constructionist and perspectivist manner. Marks for perspective constructions may be seen in many drawings of early and High Renaissance vintage.

For perspectively correct rendition, the graticulate frame, marked off in squares to facilitate proportionate enlargement or reduction, allowed the object to be drawn to be viewed in line with a screen on the drawing surface. Fixed points can be marked with relative ease on the resultant system of coordinates.

For portrait drawings, the glass board used into the 19th century had contours and important interior reference points marked on it with grease crayons or soap sticks, so that they could be transferred onto paper by tracing or direct copying. Both processes are frequently used for preliminary sketches for engravings to be duplicated, as is the screened transmission of a preliminary sketch onto the engraving plate or, magnifying, the painting surface. In such cases the screen lies over the preparatory drawing.

Mirrors and mirror arrangements with reducing convex mirrors or concave lenses were likewise used (especially in the 17th and 18th centuries) as drawing aids in the preparation of reproductions.

Even when it was a matter of the most exact rendition of topographical views, such apparatus, as well as the camera obscura (a darkened enclosure having an aperture usually provided with a lens through which light from external objects enters to form an image on the opposite surface), were frequently employed. In a darkened room the desired section is reflected through a lens onto a slanting mirror and from that inverse image is reflected again onto the horizontally positioned drawing surface. Lateral correction can be obtained by means of a second mirror.

Unless the proportions do not allow it, true-to-scale reducing or enlarging can also be carried out with the aid of the tracing instrument called the pantograph. When

ART FOR PSYCHOTHERAPY

copying, the crayon, or pencil inserted in the unequally long feet of the device reproduces the desired contours on the selected scale.

Most of these aids were thus used in normal studio practice and for the preparation of certain applied drawings. Equally practical, but useful only for closely circumscribed tasks, were elliptic compasses, curved rulers, and stencils, particularly for ornamental and decorative purposes. Only a few present-day artists, notably Jasper Johns, use stencils or simple blocks with a given shape in larger scale composition, in order to obtain the effect of repetition, often in an arbitrary use, in "alienating" technique and colour.

Mechanically produced drawings—such as typewriter sketches, computer drawings, oscillograms—and drawings done with the use of a projector, all of which can bring forth unusual and attractive results, nevertheless do not belong to the topic because they lack the immediate creativity of the art drawing.

Applied drawings

Applied and technical drawings differ in principle from art drawings in that they record unequivocally an objective set of facts and on the whole disregard aesthetic considerations. The contrast to the art drawing is sharpest in the case of technical project drawings, the purpose of which is to convey not so much visual plausibility as to give exact information that makes possible the realization of an idea.

Such plans for buildings, machines, and technical systems are not instantly readable because of the

ART FOR PSYCHOTHERAPY

orthogonal (independent) projection, the division into separate planes of projection, and the use of symbols. Prepared as a rule with such technical aids as ruler and compass, they represent a specialized language of their own, which must be learned. For topographic (detailed delineation of the features of a place) and cartographic (map-making) drawings, too, a special terminology has developed that above all systematizes spatial representations, making them intelligible to the expert with the aid of emblems and symbols.

Equally far removed from any claim to artistic standing are most illustrations serving scientific purposes, the aim of which is to record as objectively as possible the characteristic and typical features of a given phenomenon.

The systematic drawings, used especially in the natural sciences to explain a system or a function, resemble plans; descriptive and naturalistic illustrations, on the other hand, approach the illusionistic plausibility of visual experience and can attain an essentially artistic character.

A good many artists have drawn scientific illustrations, and their works—the botanical and zoological drawings of the Swiss Merian family in the 17th and 18th centuries, for example—are today more esteemed for their artistic than for their documentary value.

Of a similarly ambivalent nature is the illustrative drawing that perhaps does not go beyond a simple pictorial rendition of a literary description but because of its specific formal execution may still satisfy the highest artistic demands. Great artists have again and again illustrated Bibles, prayerbooks, novels, and literature of

ART FOR PSYCHOTHERAPY

all kinds. Some of the famous examples are Botticelli's illustrations for Dante's *Divine Comedy* and Dürer's marginal illustrations for the emperor Maximilian's prayer book.

Some artists have distinguished themselves more as illustrators than as autonomous draftsmen, as for example the 18th-century German engraver Daniel Chodowiecki, the 19th-century caricaturist Honoré Daumier, the 19th-century satiric artist Wilhelm Busch, and the 20th-century Austrian illustrator Alfred Kubin.

Clearly connected with illustrative drawing is caricature, which, by formally overemphasizing the characteristic traits of a person or situation, creates a suggestive picture that—precisely because of its distortion—engraves itself on the viewer's mind.

This special kind of drawing was done by such great artists as Leonardo, Dürer, and the 17th-century artist Gian Lorenzo Bernini and by draftsmen who, often for purposes of social criticism, have devoted themselves wholly to caricaturing, such as the 18th-century Italian Pier Leone Ghezzi, the 19th-century Frenchman Grandville (professional name of Jean-Ignace-Isidore Gérard), the 19th-century American political cartoonist Thomas Nast, and the 20th-century American Al Hirschfeld.

From such overdrawn types developed continuous picture stories that could dispense to a considerable extent with the explanatory text. Modern cartoons are based on these picture stories. Through the formally identical treatment of peculiar types, these drawings acquire an element of consecutiveness that, by telling a

ART FOR PSYCHOTHERAPY

continuing story, adds a temporal dimension to two-dimensional drawing.

This element is strongest in trick drawings that fix on paper, in brief segments of movement, invented creatures and phenomena that lack all logical plausibility; a rapid sequence of images (leafing through the pages, seeing it projected on the screen) turns the whole into apparent motion, the fundamental process of animation. The artistic achievement, if any, lies in the original invention; its actual realization is predetermined and sometimes carried out by a large and specialized staff of collaborators, often with the aid of stencils and traced designs.

Moreover, since the final result is partially determined by the mechanical multiplication, an essential criterion of drawing—the unity of work and result—does not apply.

Subject matter of drawing

Anything in the visible or imagined universe may be the theme of a drawing. In practice, however, by far the greatest number of art drawings in the Western world deal with the human figure. This situation springs from the close bond between drawing and painting: in sketches, studies, and compositions, drawing prepared the way for painting by providing preliminary clarification and some formal predetermination of the artist's concept of a given work. Many drawings now highly regarded as independent works were originally "bound," or "latent," in that they served the ends of painting or sculpture.

Yet, so rounded, self-contained, and aesthetically satisfying are these drawings that their erstwhile role as

ART FOR PSYCHOTHERAPY

handmaidens to the other pictorial arts can be reconstructed only from knowledge of the completed work, not from the drawing itself. This situation is especially true of a pictorial theme that acquired, at a relatively early stage, an autonomous rank in drawing itself: the portrait.

Portraits

Drawn 15th-century portraits—by Pisanello or Jan van Eyck, for example—may be considered completed pictorial works in their concentration, execution, and distribution of space. The clear, delicately delineated representation follows every detail of the surface, striving for realism.

The profile, rich in detail, is preferred; resembling relief, it is akin to the medallion. Next in prominence to the pure profile, the three-quarter profile, with its more spatial effect, came to the fore, to remain for centuries the classic portrait stance.

The close relationship to painting applies to practically all portrait drawings of the 15th century. Even so forceful a work as Dürer's drawing of the emperor Maximilian originated as a portrait study for a painting. At the same time, however, some of Dürer's portrait drawings clearly embody the final stage of an artistic enterprise, an ambivalence that can also be observed in other 16th-century portraitists.

The works of Jean and François Clouet in France and of the younger Hans Holbein in Switzerland and even more markedly in England in the same century bestowed an

ART FOR PSYCHOTHERAPY

autonomy on portrait drawing, especially when a drawing was completed in chalk of various colours.

The choice of the softer medium, the contouring, which for all its exactitude is less severely self-contained, and the more delicate interior drawing with plane elements gives these drawings a livelier, more personal character and accentuates once more their proximity to painting.

In polychromatic chalk technique and pastel, portrait drawing maintained its independence into the 19th century. In the 18th century, Quentin de La Tour, François Boucher, and Jean-Baptiste Chardin—all of these artists from France—were among its chief practitioners, and even Ingres, living in the 19th century, still used its technique. In pastel painting, the portrait outweighed all other subjects.

In the choice of pose, type, and execution, portrait painting, like other art forms, is influenced by the general stylistic features of an epoch. Thus, the extreme pictorial attitude of the late Baroque and Rococo was followed by a severer conception during Neoclassicism, which preferred monochrome techniques and cultivated as well the special form of the silhouette, a profile contour drawing with the area filled in in black.

Unmistakably indebted to their 15th-century predecessors, the creators of portrait drawings of the early 19th century aimed once more at the exact rendition of detail and plastic effects gained through the most carefully chosen graphic mediums: the thin, hard pencil was their favourite instrument, and the silverpoint, too, was rediscovered by the Romantics.

More interested in the psychological aspects of portraiture, late 19th- and 20th-century draftsmen prefer

ART FOR PSYCHOTHERAPY

the softer crayons that readily follow every artistic impulse. The seizing of characteristic elements and an adequate plane rendition weigh more heavily with them than realistic detail. Mood elements, intellectual tension, and personal engagement are typical features of the modern portrait and thus also of modern portrait drawing, an art that continues to document the artist's personal craftsmanship beyond the characteristics of various techniques.

Landscapes

As early as the 15th century, landscape drawings, too, attained enough autonomy so that it is hard to distinguish between the finished study for the background of a particular painting and an independent, self-contained sketched landscape.

Already in Jacopo Bellini's 15th-century sketchbooks (preserved in albums in the British Museum and the Louvre), there is an intimate connection between nature study and pictorial structure; in Titian's studio in the 16th century, landscape sketches must have been displayed as suggestions for pictorial backgrounds.

But it was Dürer who developed landscape as a recollected image and autonomous work of art, in short, as a theme of its own without reference to other works. His watercolours above all but also the drawings of his two Italian journeys, of the surroundings of Nürnberg, and of the journey to the Netherlands, represent the earliest pure landscape drawings. Centuries had to pass before such drawings occurred again in this absolute formulation.

ART FOR PSYCHOTHERAPY

Landscape elements were also very significant in 16th-century German and Dutch drawings and illustrations. The figurative representation, still extant in most cases, is formally quite integrated into the romantic forest-and-meadow landscape, particularly in the works of the Danube School—Albrecht Altdorfer and Wolf Huber, for example. More frequently than in other schools, one finds here carefully executed nature views. In the Netherlands, Pieter Bruegel drew topographical views as well as free landscape compositions, in both cases as autonomous works.

In the 17th century, the nature study and the landscape drawing that grew out of it reached a new high. The landscape drawings of the Accademia degli Incamminati (those of Domenichino, for example) combined classical and mythological themes with heroic landscapes. The Frenchman Claude Lorrain, living in Rome, frequently worked under the open sky, creating landscape drawings with a hitherto unattained atmospheric quality.

This type of cultivated and idealized landscape, depicted also by Poussin and other Northerners residing in Rome (they were called Dutch Romanists in view of the fact that so many artists from the Netherlands lived in Rome, their drawings of Italy achieving an almost ethereal quality), is in contrast with the unheroic, close-to-nature concept of landscape held primarily by the Netherlanders when depicting the landscape of their native country.

All landscape painters—their landscape paintings a specialty that was strongly represented in the artistically specialized Low Countries—also created independent landscape drawings (Jan van Goyen and Jacob van Ruisdael and his uncle and cousin, for example), with

ART FOR PSYCHOTHERAPY

Rembrandt again occupying a special position: capturing the characteristics of a region often with only a few strokes, he enhanced them in such manner that they acquire monumental expressive power even in the smallest format.

In 18th-century Italy, the topographically faithful landscape drawing gained in importance with the advent of the *Vedutisti*, the purveyors of "views," forming a group by themselves (among them, Giambattista Piranesi and Canaletto [Giovanni Antonio Canal]) and often working with such optical aids as the graticulate frame and camera obscura.

Landscape drawings of greater artistic freedom, as well as imaginary landscapes, were done most successfully by some French artists, among them Hubert Robert; pictorially and atmospherically, these themes reached a second flowering in the brush-drawn landscapes of such English artists as Turner and Alexander Cozens, whose influence extends well into the 20th century.

Given their strong interest in delineation, the 18th-century draftsmen of Neoclassicism and, even more, of Romanticism observed nature with topographical accuracy. As a new "discovery," the romantically and heroically exaggerated Alpine world now took its place in the artist's mind alongside the arcadian view of the Italian landscape.

Landscape drawings and even more, watercolours, formed an inexhaustible theme in the 19th century. The French artist Jean-Baptiste-Camille Corot and, toward the end of the century, Cézanne and van Gogh, were among the chief creators of landscape drawings. Landscapes formed part of the work of many 20th-century draftsmen,

ART FOR PSYCHOTHERAPY

but, for much of the century, the genre as such took second place to general problems of form, in which the subject was treated merely as a starting point.

However, during the last 30 years of the 20th century, a large number of American artists returned to representation, thus reinvesting in the landscape as a subject.

Figure compositions and still lifes

Compared to the main themes of autonomous drawing—portraiture and landscape—all others are of lesser importance. Figure compositions depend greatly on the painting of their time and are often directly connected with it. There were, to be sure, artists who dealt in their drawings with the themes of monumental painting, such as the 17th-century engraver and etcher Raymond de La Fage; in general, however, the artistic goal of figure composition is the picture, with the drawing representing but a useful aid and a way station.

Genre scenes, especially popular in the 17th-century Low Countries (as done by Adriaen Brouwer, Adriaen van Ostade, and Jan Steen, for example) and in 18th-century France and England, did attain some independent standing. In the 19th century, too, there were drawings that told stories of everyday life; often illustrative in character, they may be called "small pictures," not only on account of the frequently multicoloured format but also in their artistic execution.

Still lifes can also lay claim to being autonomous drawings, especially the representations of flowers, such

ART FOR PSYCHOTHERAPY

as those of the Dutch artist Jan van Huysum, which have been popular ever since the 17th century.

Here, again, it is true that a well-designed arrangement transforms an immediate nature study into a pictorial composition. In some of these compositions the similarity to painting is very strong; the pastels of the 19th- and 20th-century artist Odilon Redon, for instance, or the work of the 20th-century German Expressionist Emil Nolde, with its chromatic intensity, transcend altogether the dividing line between drawing and painting. In still lifes, as in landscapes, autonomous principles of form are more important to modern artists than the factual statement.

Fanciful and non-representational drawings

Drawings with imaginary and fanciful themes are more independent of external reality. Dream apparitions, metamorphoses, and the entwining of separate levels and regions of reality have been traditional themes. The late 15th-century phantasmagoric works of Hieronymus Bosch are an early example. There are allegorical peasant scenes by the 16th-century Flemish artist Pieter Bruegel and the carnival etchings of the 17th-century French artist Jacques Callot.

Others whose works illustrate what can be done with drawing outside landscape and portraiture are: the 18th-century Italian engraver Giambattista Piranesi, the 18th-century Anglo-Swiss artist Henry Fuseli, the 19th-century English illustrator Walter Crane, the 19th-century French Symbolist artist Gustave Moreau, and the 20th-century Surrealists.

ART FOR PSYCHOTHERAPY

Non-representational art, with its reduction of the basic elements of drawing—point, line, plane—to pure form, offered new challenges. Through renunciation of associative corporeal and spatial relationships, the unfolding of the dimensions of drawing and the structure of the various mediums acquire new significance.

The graphic qualities of the line in the plane as well as the unmarked area had already been emphasized in earlier times—for example, in the *grotteschi* of Giuseppe Arcimboldo in the 16th century (the fanciful or fantastic representations of human and animal forms often combined with each other and interwoven with representations of foliage, flowers, fruit, or the like) and in calligraphic exercises such as moresques (strongly stylized linear ornament, based on leaves and blossoms)—but mostly as printing or engraving models for the most disparate decorative tasks (interior decoration, furniture, utensils, jewelry, weapons, and the like).

Artistic architectural drawings

There is one field in which drawing fulfils a distinct function: artistic architectural drawings are a final product as drawings, differing from the impersonal, exact plans and designs by the same "handwriting" character that typifies art drawings. In many cases, no execution of these plans was envisaged; since the early Renaissance, such ideal plans have been drawn to symbolize, in execution and accessories, an abstract content.

Despite the often considerable exactitude with which the plans are drawn, the personal statement predominates in the flow of the line. This personal note clearly identifies

ART FOR PSYCHOTHERAPY

the drawings of such artists and architects as Albrecht Altdorfer, Leonardo, Michelangelo, Bernini, Francesco Borromini, and Piranesi.

Also distinct from the ground-plan type of architectural drawing are the art drawings of autonomous character created by such 20th-century architects as Erich Mendelsohn and Le Corbusier.

History of drawing - Western

As an artistic endeavour, drawing is almost as old as mankind. In an instrumental, subordinate role, it developed along with the other arts in antiquity and the Middle Ages. Whether preliminary sketches for mosaics and murals or architectural drawings and designs for statues and reliefs within the variegated artistic production of the Gothic medieval building and artistic workshop, drawing as a nonautonomous auxiliary skill was subordinate to the other arts.

Only in a very limited sense can one speak of centres of drawing in the early and High Middle Ages; that is, the scriptoria of the monasteries of Corbie and Reims in France, as well as those of Canterbury and Winchester in England, and also a few places in southern Germany, where various strongly delineatory (graphically illustrated) styles of book illumination were cultivated.

14th, 15th, and 16th centuries

In the West, the history of drawing as an independent artistic document began toward the end of the 14th century. If its development was independent, however, it was not insular. Just as the greatest draftsmen have been

ART FOR PSYCHOTHERAPY

for the most part also distinguished painters, illustrators, sculptors, or architects, so the centres and the high points of drawing have generally coincided with the leading localities and the major epochs of the other arts.

Moreover, the same stylistic phenomena have been expressed in drawing as in other art forms. Indeed, drawing shares with other art forms the characteristics of individual style, period style, and regional features. Drawing differs, however, in that it interprets and renders these characteristics in terms of its own unique mediums.

Drawing became an independent art form in northern Italy, at first quite within the framework of ordinary studio activity. But with nature studies, copies of antiques, and drafts in the various sketchbooks (those of Giovannino de'Grassi, Antonio Pisanello, and Jacopo Bellini, for example), the tradition of the Bauhütten studio workshop changed to individual work: the place of "exempla," models, reproduced in formalized fashion was now being taken by subjectively probing and partially creative drawings. In the early 15th century the international Soft Style of the period still largely predominated over the draftsman's individual "handwriting".

At mid-century, however, the differentiation of drawing style according to region and the artist's personality set in. Essential criteria, destined to remain characteristic for generations, begin to strike the eye.

In drawing produced north of the Alps, the characteristic features lie in the tendency to pictorial compactness and precise execution of detail. Many painters produced individual drawings, but the most notable draftsmen are the otherwise unidentified 15th-century German Master of the Housebook and his contemporary Martin Schongauer.

ART FOR PSYCHOTHERAPY

Both of these artists were also major copperplate engravers, so that it is not always easy to determine whether the work is a preliminary sketch or an independent drawing.

In Italian Renaissance drawings, of which there are a great many, the diverging stylistic features of the various artistic regions were particularly evident. What they had in common was the overwhelming importance of the sketch and the study, in contrast to the far rarer finished drawings. The formal and thematic connection with painting is very close even when it was not a question of preliminary drawings. The draftsmen of Venice and northern Italy preferred an open form with loose and interrupted delineation in order to achieve even in drawing the pictorial effect that corresponded to their painters' imagination.

In central Italy, on the other hand, and especially in Florence, it was the clear contour that predominated, the closed and firmly circumscribed form, the static and plastic character.

Corresponding to the functional purpose of drawing, the individual artists' studios (which, as was the case with the Medicis' Academy of St. Mark, also had to engage in general educational and humanistic investigations) formed the most significant centres of art drawing. In these large studios, drawing served not only for the probing realization of creative ideas, it was not only study and mediator between the conception and the master's finished work; it functioned also as teaching aid for the assistants who worked with the master and as a vehicle for the formation and preservation of an individual workshop tradition.

ART FOR PSYCHOTHERAPY

Although Leonardo's scientific interests were expressed in a large number of drawings, his ideal concept of the human figure is much more frequently preserved in the drawings of his collaborators and successors than in his own. Raphael and Michelangelo were also outstanding draftsmen.

Each of them used drawing in order to allow his thoughts about individual works to mature; each had a highly personal drawing style, the one with a soft and rounded stroke, the other with a sculptor's intermittent and powerful stroke. Probably a great deal of drawing was done in Raphael's studio, especially if only for the preparation of the engravings after Raphael's compositions.

From Michelangelo's hand came the first so-called connoisseur drawings that are esteemed as a personal document. They are the precursors of the collector's drawings that began in the later 16th century (autonomous works, destined for collections).

North of the Alps the autonomy of drawing was championed in the first instance by Albrecht Dürer, an indefatigable draftsman who mastered all techniques and exercised an enduring and widespread influence. The delineatory constituent clearly predominates even in his paintings.

This corresponds to the general stylistic character of 16th-century German art, within which Matthias Grünewald, with his freer, broader, and therefore more pictorial style of drawing, and the painters of the Danube school, with their ornamentalizing and agitated stroke, represent significant exceptions. In their metamorphosing of the perceived reality into drawings, the landscapes of Albrecht Altdorfer and Wolf Huber in particular are astonishing documents of a feeling for nature that might almost be called Romantic.

Soberer, incredibly compact in their pictorial concept and yet akin to the Renaissance in their objective viewing, were the portrait drawings of Hans Holbein, the Younger, whose sojourns in 16th-century England proved stimulating to other artists as well.

Similar, if less personal than Holbein because of the stricter linearity of their work, were the drawings of the French portraitists Jean and François Clouet. In the Low Countries, where they were combined with the idealized image of Italy (as in the drawings of Lucas van Leyden), Dürer's methods gained lasting popularity in the landscape drawings and studies "after life" by Pieter Bruegel the Elder.

Drawing acquired a pivotal significance in the period of Mannerism (c. 1525–1600), both as a document of artistic invention and as a means of its realization. Jacopo da Pontormo in Florence, Parmigianino in northern Italy, and Tintoretto in Venice used point and pen as essential and spontaneous vehicles of expression. Their drawings were clearly related to their painting, both in content and in the graphic method of sensitive contouring and daringly drawn foreshortening.

17th, 18th, and 19th centuries

In the early 17th century, Jacques Callot rose to prominence in French art: gifted as a draftsman above all,

ART FOR PSYCHOTHERAPY

he recorded with the pen his clever inventions and great picture stories, primarily in bold abbreviations.

The importance of drawing for an artist's growth and the widening of his horizon is attested also by the work of Peter Paul Rubens, whose studies and sketches make up an integral part of his creative achievement. In order to disseminate his pictorial themes and concept of form, he maintained his own school for draftsmen and engravers. Among the circle of Flemings around him, Jacob Jordaens and Sir Anthony Van Dyck are notable as draftsmen with a style of their own.

Hercules Seghers was among the most fascinating artists of the 17th century, a creator of drawn and etched landscapes that he continued to rework while experimenting with printing processes. From the point of view of technique and form, he was important for the greatest artist of Holland, Rembrandt. Seghers combined great inventiveness, especially in his interpretations of Old Testament motifs, and broad mastery of all the techniques of drawing. In his studio, too, drawing was emphasized as a teaching aid and a means of formal experimentation.

Most Dutch painters of the 17th century, such as the van de Velde family, Brouwer, van Ostade, Pieter Saenredam, and Paulus Potter, were also industrious draftsmen who recorded their special thematic concerns in drawings that were largely completed. Beyond serving as preparation for paintings, these were regarded as autonomous works representing the final stage of the creative process.

In 17th-century Italy, drawing by way of artistic practice and experimentation became established in the academies, especially in Bologna. More significant,

ART FOR PSYCHOTHERAPY

however, was the continuing development of landscape drawing, as initiated by the brothers Agostino and Annibale Carracci and articulated further by Domenichino and Salvator Rosa.

The French artist Claude Lorrain so developed the landscape drawing of the Roman countryside that it became almost a genre of its own; in his works, which were often intended for sale, nature study and an idealized pictorial concept are uniquely merged. In detailed studies directly before the object, he achieved a timeless validity. Like Lorrain, Nicolas Poussin also drew under the open sky.

Using various techniques, he combined realistic experiences and humanistic concepts in idealizing compositions the figures and scenes of which are harmoniously integrated into a spacious landscape. This open-air painting and drawing was practiced also by some other artists who spent a considerable time in Rome—the Dutch artists Jan Asselijn, Claes Berchem, Karel Dujardin, and Adam Pijnacker, for example. For most southern European artists of the 17th century, however, drawing was a mere stage in the creation of a painting.

Antoine Watteau, too, did drawings to "keep his hand in" for his painting, although he did so with an independence that led him far beyond the immediate occasion. Most figures in the paintings from various periods of his career were based on earlier drawings. In the grand scale of his form and the attention paid to pictorial elements, he carried on in the manner of Rubens, combining it with the light esprit of the 18th century.

ART FOR PSYCHOTHERAPY

The leading position of French art in the first half of that century was confirmed by the achievements of François Boucher, Jean-Honoré Fragonard, Hubert Robert, and Gabriel de Saint-Aubin, whose drawings include figure studies, genre-like works, and landscapes.

In contrast to the French draftsmen who brought about a flowering of the *à trois crayons* method on tinted paper, some artists created similar landscapes with pen and brush but with greater objective abbreviation.

Mention must here be made of Venice, with the Giovanni Battista Tiepolo family, whose expansively conceived pen drawings, washed with a broad brush, call forth the kind of luminaristic effect that Francesco Guardi also used for landscape studies and imaginary scenes. These had been preceded by Canaletto's views of Venice, composed more severely as far as tectonic (constructional) detail is concerned but nonetheless the first examples of this form of the landscape capriccio, or fantasy.

The architect Giovanni Battista Piranesi made his name primarily as a draftsman who recorded views of Rome; above all, in his drawings of architecture and eerie vaults (*Carceri*), he left behind a body of work of great intellectual and formal forcefulness.

The Spanish painter Francisco de Goya, at the very end of the 18th and in the beginning of the 19th century, was in advance of his time in the way in which he handled his themes. Forming an odd contrast to the court-painter's pictures, his brush-and-sanguine drawings are rather more closely tied to his cycles of etchings.

He combined the luminaristic effects of Tiepolo's drawings with the dramatic impact of a Rembrandt chiaroscuro.

ART FOR PSYCHOTHERAPY

Also at the turn of the 19th century is an artist whose main work was that of a draftsman: the English caricaturist and social satirist Thomas Rowlandson, who produced colourful and distinctive watercolours. The late 18th and, even more, the early 19th century produced a drawing style that, in accordance with both the Neoclassical and the Romantic ideal, emphasized once more the linear element.

In Jean-Auguste-Dominique Ingres, idealistic Neoclassicism found an exemplary expression of strict linearity, and the pencil drawing became a downright classical form. The Nazarenes and Romantics in Rome and the Alpine region (Joseph Anton Koch, the brothers Friedrich and Ferdinand Olivier, and Julius Schnorr von Carolsfeld) as well as those in north Germany (Philipp Otto Runge and Caspar David Friedrich) were more lyrical but equally rigorous in the use of the hardpoint; after a long time, they were the first northern artists to have made a significant contribution to the history of drawing.

Among 19th-century artists, the emphasis on delineation was characteristic also of Moritz von Schwind in Germany and John Millais in England. (In the Neoclassical phase of the 20th century it was renewed, in a more open and "handwriting" fashion, by Thomas Eakins in the United States as well as by Picasso, Matisse, and Amedeo Modigliani in France.)

The drawings of Eugène Delacroix, while preserving plastic qualities, show a broader stroke and are thus more pictorial. Honoré Daumier, active in all mediums primarily as a draftsman, utilized pictorial chiaroscuro effects in forcible statements of social criticism.

ART FOR PSYCHOTHERAPY

France continued to be a leading centre of the art of drawing, a form that was given a very personal note in each case in the works of Edgar Degas, Henri de Toulouse-Lautrec, Vincent van Gogh, and Paul Cézanne. The line—the common point of departure for all of the above-mentioned artists—did not disappear until Georges Seurat's plane shading, done in the Pointillist manner.

Modern

Except for a few stylistic currents such as Tachism (paintings consisting of irregular blobs of colour), drawing is represented in the work of practically all 20th-century artists; it is as international as modern art itself. As the other arts have become non-representational, thus attaining autonomy and formal independence in relation to external reality, drawing is more than ever considered an autonomous work of art, independent of the other arts.

Some schools and individual artists as well have concentrated on drawing and in very individualistic ways. The German Expressionists, for instance, developed especially emphatic forms of drawing with powerful delineation and forcible and hyperbolic formal description; notable examples are the works of Ernst Barlach, Käthe Kollwitz, Alfred Kubin, Ernest Ludwig Kirchner, Karl Schmidt-Rottluff, Max Beckmann, and George Grosz. In the artists' group Der Blaue Reiter (The Blue Rider), Wassily Kandinsky was foremost in laying the groundwork for a new evaluation of the non-representational line.

Paul Klee's lyrically sensitive drawings, carried out in a pen technique of unheard-of sublimity, represent a high point of modern drawing. In France, drawing plays a

ART FOR PSYCHOTHERAPY

major role, especially in the work of the painters of the École de Paris (School of Paris), such as Pierre Soulages and Hans Hartung, who consider the line, the framework of lines, and the network of lines, as primary manifestations of form. Wols (Alfred Otto Wolfgang Schulze) and also the English artist Graham Sutherland may actually be called spiritual draftsmen who put their faith in the magic of the line.

Finally, drawing occupies a considerable place in the work (including all its variants of style and form) of Pablo Picasso, who knew how to make use of its manifold technical possibilities. One is surely justified in calling him the greatest draftsman of the 20th century and one of the greatest in the history of drawing.

Eastern

Some form of monochromatic brush drawing with ink may have been practiced in China as early as the 2nd millennium BC; but the earliest pictorial work is in lacquer or on bronze vessels, contemporaneous with Alexander the Great (ruled 336–323 BC). It relies on contour and silhouette, with men and animals depicted in horizontal registers (levels, one above the other) reminiscent of Egyptian and Mediterranean work.

The extent of any mutual influence between East and West cannot yet be determined. Under the Eastern Han dynasty (AD 25–220) wall paintings, linear in character, were produced in fresco (wet plaster) and secco (dry). Only in the Wei (386–534/35) and Tang (618–907) dynasties did the true character of Chinese drawing on silk or paper emerge. In the 7th century, the characteristic albums (*ceye*) of drawings appear.

ART FOR PSYCHOTHERAPY

No distinction was made between drawing and painting because all Chinese pictorial art was fundamentally graphic. The artist worked with the fine point of the brush on paper or silk laid horizontally on a table. Work in pure outline was called *baimiao*; ink applied in splashes, *pomo*. Colour was used sparingly or not at all. The final work was not made direct from nature.

Hindu and Buddhist paintings at Ajanta in India and also in Sri Lanka reveal the essential quality in all Indian art: emphasis on a flowing, rhythmic contour to express movement and gesture. Drawings on palm leaf of the 11th century are similarly based on the use of line to depict mythological scenes.

The 14th century saw the manufacture of paper, introduced from China, permitting the production of the vertical book. Despite the Muslim prohibition of human representation, books illustrated with drawings, sometimes with flat decorative colour, were produced at the Persian and Mughal courts, but not for public display.

The use of a precise and expressive line constituted the basis for Persian and Indian (both Mughal and Rajput) miniature paintings, which show people in landscape or in relation to buildings.

Japanese art tended to follow that of China until the early 19th century, when the popular colour print was introduced. In the graceful feminine gestures of Utamaro's work, the Eastern love of flowing contour is manifest, his lines varying in width and density.

Hokusai's drawings of social life in a humorous, almost grotesque vein reveal his complete command of the expressive line.

ART FOR PSYCHOTHERAPY

SECTION FOUR: SCULPTURE

Sculpture as an art

Sculpture is an artistic form in which hard or plastic materials are worked into three-dimensional art objects. The designs may be embodied in freestanding objects, in reliefs on surfaces, or in environments ranging from tableaux to contexts that envelop the spectator. An enormous variety of media may be used, including clay, wax, stone, metal, fabric, glass, wood, plaster, rubber, and random "found" objects. Materials may be carved, modelled, moulded, cast, wrought, welded, sewn, assembled, or otherwise shaped and combined.

Sculpture is not a fixed term that applies to a permanently circumscribed category of objects or sets of activities. It is, rather, the name of an art that grows and changes and is continually extending the range of its activities and evolving new kinds of objects. The scope of the term was much wider in the second half of the 20th century than it had been only two or three decades before, and in the fluid state of the visual arts at the turn of the 21st century nobody can predict what its future extensions are likely to be.

Certain features which in previous centuries were considered essential to the art of sculpture are not present in a great deal of modern sculpture and can no longer form part of its definition. One of the most important of these is representation. Before the 20th century, sculpture was considered a representational art, one that imitated forms in life, most often human figures but also inanimate objects, such as game, utensils, and

ART FOR PSYCHOTHERAPY

books. Since the turn of the 20th century, however, sculpture has also included non-representational forms.

It has long been accepted that the forms of such functional three-dimensional objects as furniture, pots, and buildings may be expressive and beautiful without being in any way representational; but it was only in the 20th century that non-functional, non-representational, three-dimensional works of art began to be produced.

Before the 20th century, sculpture was considered primarily an art of solid form, or mass. It is true that the negative elements of sculpture—the voids and hollows within and between its solid forms—have always been to some extent an integral part of its design, but their role was a secondary one. In a great deal of modern sculpture, however, the focus of attention has shifted, and the spatial aspects have become dominant. Spatial sculpture is now a generally accepted branch of the art of sculpture.

It was also taken for granted in the sculpture of the past that its components were of a constant shape and size and, with the exception of items such as Augustus Saint-Gaudens's *Diana* (a monumental weather vane), did not move. With the recent development of kinetic sculpture, neither the immobility nor immutability of its form can any longer be considered essential to the art of sculpture.

Finally, sculpture since the 20th century has not been confined to the two traditional forming processes of carving and modelling or to such traditional natural materials as stone, metal, wood, ivory, bone, and clay. Because present-day sculptors use any materials and methods of manufacture that will serve their purposes,

ART FOR PSYCHOTHERAPY

the art of sculpture can no longer be identified with any special materials or techniques.

Through all these changes, there is probably only one thing that has remained constant in the art of sculpture, and it is this that emerges as the central and abiding concern of sculptors: the art of sculpture is the branch of the visual arts that is especially concerned with the creation of form in three dimensions.

Sculpture may be either in the round or in relief. A sculpture in the round is a separate, detached object in its own right, leading the same kind of independent existence in space as a human body or a chair. A relief does not have this kind of independence. It projects from and is attached to or is an integral part of something else that serves either as a background against which it is set or a matrix from which it emerges.

The actual three-dimensionality of sculpture in the round limits its scope in certain respects in comparison with the scope of painting. Sculpture cannot conjure the illusion of space by purely optical means or invest its forms with atmosphere and light as painting can. It does have a kind of reality, a vivid physical presence that is denied to the pictorial arts.

The forms of sculpture are tangible as well as visible, and they can appeal strongly and directly to both tactile and visual sensibilities. Even the visually impaired, including those who are congenitally blind, can produce and appreciate certain kinds of sculpture. It was, in fact, argued by the 20th-century art critic Sir Herbert Read that sculpture should be regarded as primarily an art of touch and that the roots of sculptural sensibility can be traced to the pleasure one experiences in fondling things.

ART FOR PSYCHOTHERAPY

All three-dimensional forms are perceived as having an expressive character as well as purely geometric properties. They strike the observer as delicate, aggressive, flowing, taut, relaxed, dynamic, soft, and so on. By exploiting the expressive qualities of form, a sculptor is able to create images in which subject matter and expressiveness of form are mutually reinforcing. Such images go beyond the mere presentation of fact and communicate a wide range of subtle and powerful feelings.

The aesthetic raw material of sculpture is, so to speak, the whole realm of expressive three-dimensional form. A sculpture may draw upon what already exists in the endless variety of natural and man-made form, or it may be an art of pure invention. It has been used to express a vast range of human emotions and feelings from the most tender and delicate to the most violent and ecstatic.

All human beings, intimately involved from birth with the world of three-dimensional form, learn something of its structural and expressive properties and develop emotional responses to them. This combination of understanding and sensitive response, often called a sense of form, can be cultivated and refined. It is to this sense of form that the art of sculpture primarily appeals.

Elements and principles of sculptural design

The two most important elements of sculpture—mass and space—are, of course, separable only in thought. All sculpture is made of a material substance that has mass and exists in three-dimensional space. The mass of sculpture is thus the solid, material, space-occupying bulk that is contained within its surfaces.

ART FOR PSYCHOTHERAPY

Space enters into the design of sculpture in three main ways: the material components of the sculpture extend into or move through space; they may enclose or enfold space, thus creating hollows and voids within the sculpture; and they may relate one to another across space. Volume, surface, light and shade, and colour are supporting elements of sculpture.

Elements of design

The amount of importance attached to either mass or space in the design of sculpture varies considerably. In Egyptian sculpture and in most of the sculpture of the 20th-century artist Constantin Brancusi, for example, mass is paramount, and most of the sculptor's thought was devoted to shaping a lump of solid material. In 20th-century works by Antoine Pevsner or Naum Gabo, on the other hand, mass is reduced to a minimum, consisting only of transparent sheets of plastic or thin metal rods.

The solid form of the components themselves is of little importance; their main function is to create movement through space and to enclose space. In works by such 20th-century sculptors as Henry Moore and Barbara Hepworth, the elements of space and mass are treated as more or less equal partners.

It is not possible to see the whole of a fully three-dimensional form at once. The observer can only see the whole of it if he turns it around or goes around it himself. For this reason it is sometimes mistakenly assumed that sculpture must be designed primarily to present a series of satisfactory projective views and that this multiplicity of views constitutes the main difference between

ART FOR PSYCHOTHERAPY

sculpture and the pictorial arts, which present only one view of their subject.

Such an attitude toward sculpture ignores the fact that it is possible to apprehend solid forms as volumes, to conceive an idea of them in the round from any one aspect. A great deal of sculpture is designed to be apprehended primarily as volume.

A single volume is the fundamental unit of three-dimensional solid form that can be conceived in the round. Some sculptures consist of only one volume, others are configurations of a number of volumes. The human figure is often treated by sculptors as a configuration of volumes, each of which corresponds to a major part of the body, such as the head, neck, thorax, and thigh.

Holes and cavities in sculpture, which are as carefully shaped as the solid forms and are of equal importance to the overall design, are sometimes referred to as negative volumes.

The surfaces of sculpture are in fact all that one actually sees. It is from their inflections that one makes inferences about the internal structure of the sculpture. A surface has, so to speak, two aspects: it contains and defines the internal structure of the masses of the sculpture, and it is the part of the sculpture that enters into relations with external space.

The expressive character of different kinds of surfaces is of the utmost importance in sculpture. Double-curved convex surfaces suggest fullness, containment, enclosure, the outward pressure of internal forces. In the aesthetics of Indian sculpture such surfaces have a special metaphysical significance. Representing the

ART FOR PSYCHOTHERAPY

encroachment of space into the mass of the sculpture, concave surfaces suggest the action of external forces and are often indicative of collapse or erosion.

Flat surfaces tend to convey a feeling of material hardness and rigidity; they are unbending or unyielding, unaffected by either internal or external pressures. Surfaces that are convex in one curvature and concave in the other can suggest the operation of internal pressures and at the same time receptivity to the influence of external forces. They are associated with growth, with expansion into space.

Unlike the painter, who creates light effects within the work, the sculptor manipulates actual light on the work. The distribution of light and shade over the forms of his work depends upon the direction and intensity of light from external sources. Nevertheless, to some extent he can determine the kinds of effect this external light will have. If he knows where the work is to be sited, he can adapt it to the kind of light it is likely to receive.

The brilliant overhead sunlight of Egypt and India demands a different treatment from the dim interior light of a northern medieval cathedral. Then again, it is possible to create effects of light and shade, or chiaroscuro, by cutting or modelling deep, shadow-catching hollows and prominent, highlighted ridges.

Many late Gothic sculptors used light and shade as a powerful expressive feature of their work, aiming at a mysterious obscurity, with forms broken by shadow emerging from a dark background. Greek, Indian, and most Italian Renaissance sculptors shaped the forms of their work to receive light in a way that makes the whole work radiantly clear.

ART FOR PSYCHOTHERAPY

The colouring of sculpture may be either natural or applied. In the recent past, sculptors became more aware than ever before of the inherent beauty of sculptural materials. Under the slogan of "truth to materials" many of them worked their materials in ways that exploited their natural properties, including colour and texture. More recently, however, there has been a growing tendency to use bright artificial colouring as an important element in the design of sculpture.

In the ancient world and during the Middle Ages almost all sculpture was artificially coloured, usually in a bold and decorative rather than a naturalistic manner. The sculptured portal of a cathedral, for example, would be coloured and gilded with all the brilliance of a contemporary illuminated manuscript.

Combinations of differently coloured materials, such as the ivory and gold of some Greek sculpture, were not unknown before the 17th century; but the early Baroque sculptor Gian Lorenzo Bernini greatly extended the practice by combining variously coloured marbles with white marble and gilt bronze.

Principles of design

It is doubtful whether any principles of design are universal in the art of sculpture, for the principles that govern the organization of the elements of sculpture into expressive compositions differ from style to style. In fact, distinctions made among the major styles of sculpture are largely based on a recognition of differences in the principles of design that underlie them.

ART FOR PSYCHOTHERAPY

Thus, the art historian Erwin Panofsky was attempting to define a difference of principle in the design of Romanesque and Gothic sculpture when he stated that the forms of Romanesque were conceived as projections from a plane outside themselves, while those of Gothic were conceived as being centred on an axis within themselves.

The "principle of axiality" was considered by Panofsky to be "the essential principle of classical statuary," which Gothic had rediscovered.

The principles of sculptural design govern the approaches of sculptors to such fundamental matters as orientation, proportion, scale, articulation, and balance.

For conceiving and describing the orientation of the forms of sculpture in relation to each other, to a spectator, and to their surroundings, some kind of spatial scheme of reference is required. This is provided by a system of axes and planes of reference.

An axis is an imaginary centre line through a symmetrical or near symmetrical volume or group of volumes that suggests the gravitational pivot of the mass. Thus, all the main components of the human body have axes of their own, while an upright figure has a single vertical axis running through its entire length. Volumes may rotate or tilt on their axes.

Planes of reference are imaginary planes to which the movements, positions, and directions of volumes, axes, and surfaces may be referred. The principal planes of reference are the frontal, the horizontal, and the two profile planes.

ART FOR PSYCHOTHERAPY

The principles that govern the characteristic poses and spatial compositions of upright figures in different styles of sculpture are formulated with reference to axes and the four cardinal planes: for example, the principle of axiality already referred to; the principle of frontality, which governs the design of Archaic sculpture; the characteristic contrapposto (pose in which parts of the body, such as upper and lower, tilt or even twist in opposite directions) of Michelangelo's figures; and in standing Greek sculpture of the Classical period the frequently used balanced "chiastic" pose (stance in which the body weight is taken principally on one leg, thereby creating a contrast of tension and relaxation between the opposite sides of a figure).

Proportional relations exist among linear dimensions, areas, and volumes and masses. All three types of proportion coexist and interact in sculpture, contributing to its expressiveness and beauty. Attitudes toward proportion differ considerably among sculptors. Some sculptors, both abstract and figurative, use mathematical systems of proportion; for example, the refinement and idealization of natural human proportions was a major preoccupation of Greek sculptors.

Indian sculptors employed iconometric canons, or systems of carefully related proportions, that determined the proportions of all significant dimensions of the human figure. African and other tribal sculptors base the proportions of their figures on the subjective importance of the parts of the body. Unnatural proportions may be used for expressive purposes or to accommodate a sculpture to its surroundings. The elongation of the figures on the Portail Royal ("Royal Portal") of Chartres

ART FOR PSYCHOTHERAPY

cathedral does both: it enhances their otherworldliness and also integrates them with the columnar architecture.

Sometimes it is necessary to adapt the proportions of sculpture to suit its position in relation to a viewer. A figure sited high on a building, for example, is usually made larger in its upper parts in order to counteract the effects of foreshortening. This should be allowed for when a sculpture intended for such a position is exhibited on eye level in a museum.

The scale of sculpture must sometimes be considered in relation to the scale of its surroundings. When it is one element in a larger complex, such as the facade of a building, it must be in scale with the rest.

Another important consideration that sculptors must take into account when designing outdoor sculpture is the tendency of sculpture in the open air—particularly when viewed against the sky—to appear less massive than it does in a studio. Because one tends to relate the scale of sculpture to one's own human physical dimensions, the emotional impact of a colossal figure and a small figurine are quite different.

In ancient and medieval sculpture the relative scale of the figures in a composition is often determined by their importance; e.g., slaves are much smaller than kings or nobles. This is sometimes known as hierarchic scale.

The joining of one form to another may be accomplished in a variety of ways. In much of the work of the 19th-century French sculptor Auguste Rodin, there are no clear boundaries, and one form is merged with another in an impressionistic manner to create a continuously flowing surface. In works by the Greek sculptor

ART FOR PSYCHOTHERAPY

Praxiteles, the forms are softly and subtly blended by means of smooth, blurred transitions.

The volumes of Indian sculpture and the surface anatomy of male figures in the style of the Greek sculptor Polyclitus are sharply defined and clearly articulated. One of the main distinctions between the work of Italian and northern Renaissance sculptors lies in the Italians' preference for compositions made up of clearly articulated, distinct units of form and the tendency of the northern Europeans to subordinate the individual parts to the all over flow of the composition.

The balance, or equilibrium, of freestanding sculpture has three aspects. First, the sculpture must have actual physical stability. This can be achieved by natural balance—that is, by making the sculpture stable enough in itself to stand firmly—which is easy enough to do with a four-legged animal or a reclining figure but not with a standing figure or a tall, thin sculpture, which must be secured to a base.

The second aspect of balance is compositional. The interaction of forces and the distribution of weight within a composition may produce a state of either dynamic or static equilibrium. The third aspect of balance applies only to sculpture that represents a living figure. A live human figure balances on two feet by making constant movements and muscular adjustments. Such an effect can be conveyed in sculpture by subtle displacements of form and suggestions of tension and relaxation.

ART FOR PSYCHOTHERAPY

Relationships to other arts

Sculpture has long been closely related to architecture through its role as architectural decoration and also at the level of design. Architecture, like sculpture, is concerned with three-dimensional form; and, although the central problem in the design of buildings is the organization of space rather than mass, there are styles of architecture that are effective largely through the quality and organization of their solid forms. Ancient styles of stone architecture, particularly Egyptian, Greek, and Mexican, tend to treat their components in a sculptural manner. Moreover, most buildings viewed from the outside are compositions of masses.

The growth of spatial sculpture is so intimately related to the opening up and lightening of architecture, which the development of modern building technology has made possible, that many 20th-century sculptors can be said to have treated their work in an architectural manner.

Some forms of relief sculpture approach very closely the pictorial arts of painting, drawing, engraving, and so on. And sculptures in the round that make use of chiaroscuro and that are conceived primarily as pictorial views rather than as compositions in the round are said to be "painterly"; for example, Bernini's *Ecstasy of St. Teresa* (Santa Maria della Vittoria, Rome).

The borderlines between sculpture and pottery and the metalworking arts are not clear-cut, and many pottery and metal artefacts have every claim to be considered as sculpture. Today there is a growing affinity between the work of industrial designers and sculptors. Sculptural modelling techniques, and sometimes sculptors themselves, are often involved, for example, in the initial stages of the design of new automobile bodies.

The close relationships that exist between sculpture and the other visual arts are attested by the number of artists who have readily turned from one art to another; for example, Michelangelo, Bernini, Pisanello, Degas, and Picasso.

Materials

Any material that can be shaped in three dimensions can be used sculpturally. Certain materials, by virtue of their structural and aesthetic properties and their availability, have proved especially suitable. The most important of these are stone, wood, metal, clay, ivory, and plaster. There are also a number of materials of secondary importance and many that have only recently come into use.

Primary

Throughout history, stone has been the principal material of monumental sculpture. There are practical reasons for this: many types of stone are highly resistant to the weather and therefore suitable for external use; stone is available in all parts of the world and can be obtained in large blocks; many stones have a fairly homogeneous texture and a uniform hardness that make them suitable for carving; stone has been the chief material used for the monumental architecture with which so much sculpture has been associated.

Stones belonging to all three main categories of rock formation have been used in sculpture. Igneous rocks,

ART FOR PSYCHOTHERAPY

which are formed by the cooling of molten masses of mineral as they approach the Earth's surface, include granite, diorite, basalt, and obsidian. These are some of the hardest stones used for sculpture. Sedimentary rocks, which include sandstones and lime stones, are formed from accumulated deposits of mineral and organic substances.

Sandstones are agglomerations of particles of eroded stone held together by a cementing substance. Lime stones are formed chiefly from the calcareous remains of organisms.

Alabaster (gypsum), also a sedimentary rock, is a chemical deposit. Many varieties of sandstone and limestone, which vary greatly in quality and suitability for carving, are used for sculpture. Because of their method of formation, many sedimentary rocks have pronounced strata and are rich in fossils.

Metamorphic rocks result from changes brought about in the structure of sedimentary and igneous rocks by extreme pressure or heat. The most well-known metamorphic rocks used in sculpture are the marbles, which are recrystallized lime stones.

Italian Carrara marble, the best known, was used by Roman and Renaissance sculptors, especially Michelangelo, and is still widely used. The best-known varieties used by Greek sculptors, with whom marble was more popular than any other stone, are Pentelic—from which the Parthenon and its sculpture are made—and Parian.

Because stone is extremely heavy and lacks tensile strength, it is easily fractured if carved too thinly and not properly supported. A massive treatment without

ART FOR PSYCHOTHERAPY

vulnerable projections, as in Egyptian and pre-Columbian American Indian sculpture, is therefore usually preferred.

Some stones, however, can be treated more freely and openly; marble in particular has been treated by some European sculptors with almost the same freedom as bronze, but such displays of virtuosity are achieved by overcoming rather than submitting to the properties of the material itself.

The colours and textures of stone are among its most delightful properties. Some stones are fine-grained and can be carved with delicate detail and finished with a high polish; others are coarse-grained and demand a broader treatment. Pure white Carrara marble, which has a translucent quality, seems to glow and responds to light in a delicate, subtle manner.

(These properties of marble were brilliantly exploited by 15th-century Italian sculptors such as Donatello and Desiderio da Settignano.)

The colouring of granite is not uniform but has a salt-and-pepper quality and may glint with mica and quartz crystals. It may be predominantly black or white or a variety of greys, pinks, and reds. Sandstones vary in texture and are often warmly coloured in a range of buffs, pinks, and reds. Lime stones vary greatly in colour, and the presence of fossils may add to the interest of their surfaces. A number of stones are richly variegated in colour by the irregular veining that runs through them.

Hard stones, or semiprecious stones, constitute a special group, which includes some of the most beautiful and decorative of all substances. The working of these stones, along with the working of more precious gemstones, is usually considered as part of the glyptic

ART FOR PSYCHOTHERAPY

(gem carving or engraving), or lapidary, arts, but many artefacts produced from them can be considered small-scale sculpture.

They are often harder to work than steel. First among the hard stones used for sculpture is jade, which was venerated by the ancient Chinese, who worked it, together with other hard stones, with extreme skill. It was also used sculpturally by Maya and Mexican artists. Other important hard stones are rock crystal, rose quartz, amethyst, agate, and jasper.

The principal material of tribal sculpture in Africa, Oceania, and North America, wood has also been used by every great civilization; it was used extensively during the Middle Ages, for example, especially in Germany and central Europe. Among modern sculptors who have used wood for important works are Ernst Barlach, Ossip Zadkine, and Henry Moore.

Both hardwoods and softwoods are used for sculpture. Some are close-grained, and they cut like cheese; others are open-grained and stringy. The fibrous structure of wood gives it considerable tensile strength, so that it may be carved thinly and with greater freedom than stone. For large or complex open compositions, a number of pieces of wood may be jointed.

Wood is used mainly for indoor sculpture, for it is not as tough or durable as stone; changes of humidity and temperature may cause it to split, and it is subject to attack by insects and fungus. The grain of wood is one of its most attractive features, giving variety of pattern and texture to its surfaces. Its colours, too, are subtle and varied. In general, wood has warmth that stone does not have, but it lacks the massive dignity and weight of stone.

ART FOR PSYCHOTHERAPY

The principal woods for sculpture are oak, mahogany, lime wood, walnut, elm, pine, cedar, boxwood, pear, and ebony; but many others are also used. The sizes of wood available are limited by the sizes of trees; North American Indians, for example, could carve gigantic totem poles in pine, but boxwood is available only in small pieces.

In the 20th century, wood was used by many sculptors as a medium for construction as well as for carving. Laminated timbers, chipboards, and timber in block and plank form can be glued, jointed, screwed, or bolted together, and given a variety of finishes.

Wherever metal technologies have been developed, metals have been used for sculpture. The amount of metal sculpture that has survived from the ancient world does not properly reflect the extent to which it was used, for vast quantities have been plundered and melted down. Countless Far Eastern and Greek metal sculptures have been lost in this way, as has almost all the goldwork of pre-Columbian American Indians.

The metal most used for sculpture is bronze, which is basically an alloy of copper and tin; but gold, silver, aluminum, copper, brass, lead, and iron have also been widely used. Most metals are extremely strong, hard, and durable, with a tensile strength that permits a much greater freedom of design than is possible in either stone or wood.

A life-size bronze figure that is firmly attached to a base needs no support other than its own feet and may even be poised on one foot. Considerable attenuation of form is also possible without risk of fracture.

The colour, brilliant lustre, and reflectivity of metal surfaces have been highly valued and made full use of in

ART FOR PSYCHOTHERAPY

sculpture although, since the Renaissance, artificial patinas have generally been preferred as finishes for bronze.

Metals can be worked in a variety of ways in order to produce sculpture. They can be cast—that is, melted and poured into moulds; squeezed under pressure into dies, as in coin making; or worked directly—for example, by hammering, bending, cutting, welding, and repoussé (hammered or pressed in relief).

Important traditions of bronze sculpture are Greek, Roman, Indian (especially Chola), African (Bini and Yoruba), Italian Renaissance, and Chinese. Gold was used to great effect for small-scale works in pre-Columbian America and medieval Europe.

A fairly recent discovery, aluminum has been used a great deal by modern sculptors. Iron has not been used much as a casting material, but in recent years it has become a popular material for direct working by techniques similar to those of the blacksmith. Sheet metal is one of the principal materials used nowadays for constructional sculpture. Stainless steel in sheet form has been used effectively by the American sculptor David Smith.

Clay is one of the most common and easily obtainable of all materials. Used for modelling animal and human figures long before men discovered how to fire pots, it has been one of the sculptor's chief materials ever since.

Clay has four properties that account for its widespread use: when moist, it is one of the most plastic of all substances, easily modelled and capable of registering the most detailed impressions; when partially dried out to a leather-hard state or completely dried, it can be carved

ART FOR PSYCHOTHERAPY

and scraped; when mixed with enough water, it becomes a creamy liquid known as slip, which may be poured into moulds and allowed to dry; when fired to temperatures of between 700 and 1,400 °C (1,300 and 2,600 °F), it undergoes irreversible structural changes that make it permanently hard and extremely durable.

Sculptors use clay as a material for working out ideas; for preliminary models that are subsequently cast in such materials as plaster, metal, and concrete or carved in stone; and for pottery sculpture.

Depending on the nature of the clay body itself and the temperature at which it is fired, a finished pottery product is said to be earthenware, which is opaque, relatively soft, and porous; stoneware, which is hard, nonporous, and more or less vitrified; or porcelain, which is fine-textured, vitrified, and translucent.

All three types of pottery are used for sculpture. Sculpture made in low-fired clays, particularly buff and red clays, is known as terra-cotta (baked earth). This term is used inconsistently, however, and is often extended to cover all forms of pottery sculpture.

Unglazed clay bodies can be smooth or coarse in texture and may be coloured white, grey, buff, brown, pink, or red. Pottery sculpture can be decorated with any of the techniques invented by potters and coated with a variety of beautiful glazes.

Palaeolithic sculptors produced relief and in-the-round work in unfired clay. The ancient Chinese, particularly during the Tang (618–907) and Song (960–1279) dynasties, made superb pottery sculpture, including large-scale human figures.

ART FOR PSYCHOTHERAPY

The best-known Greek works are the intimate small-scale figures and groups from Tanagra. Mexican and Maya sculptor-potters produced vigorous, directly modeled figures. During the Renaissance, pottery was used in Italy for major sculptural projects, including the large-scale glazed and coloured sculptures of Luca della Robbia and his family, which are among the finest works in the medium.

One of the most popular uses of the pottery medium has been for the manufacture of figurines—at Staffordshire, Meissen, and Sèvres, for example.

The main source of ivory is elephant tusks; but walrus, hippopotamus, narwhal (an Arctic aquatic animal), and, in Palaeolithic times, mammoth tusks also were used for sculpture. Ivory is dense, hard, and difficult to work. Its colour is creamy white, which usually yellows with age; and it will take a high polish.

A tusk may be sawed into panels for relief carving or into blocks for carving in the round; or the shape of the tusk itself may be used. The physical properties of the material invite the most delicate, detailed carving, and displays of virtuosity are common.

Ivory was used extensively in antiquity in the Middle and Far East and the Mediterranean. An almost unbroken Christian tradition of ivory carving reaches from Rome and Byzantium to the end of the Middle Ages. Throughout this time, ivory was used mainly in relief, often in conjunction with precious metals, enamels, and precious stones to produce the most splendid effects.

Some of its main sculptural uses were for devotional diptychs, portable altars, book covers, retables (raised shelves above altars), caskets, and crucifixes. The

ART FOR PSYCHOTHERAPY

Baroque period, too, is rich in ivories, especially in Germany. A fine tradition of ivory carving also existed in Benin, a former kingdom of West Africa.

Related to ivory, horn and bone have been used since Paleolithic times for small-scale sculpture. Reindeer horn and walrus tusks were two of the Eskimo carver's most important materials. One of the finest of all medieval "ivories" is a carving in whalebone, *The Adoration of the Magi* (Victoria and Albert Museum, London).

Plaster of Paris (sulphate of lime) is especially useful for the production of moulds, casts, and preliminary models. It was used by Egyptian and Greek sculptors as a casting medium and is today the most versatile material in the sculptor's workshop.

When mixed with water, plaster will in a short time recrystallize, or set—that is, become hard and inert—and its volume will increase slightly. When set, it is relatively fragile and lacking in character and is therefore of limited use for finished work.

Plaster can be poured as a liquid, modeled directly when of a suitable consistency, or easily carved after it has set. Other materials can be added to it to retard its setting, to increase its hardness or resistance to heat, to change its colour, or to reinforce it.

The main sculptural use of plaster in the past was for moulding and casting clay models as a stage in the production of cast metal sculpture. Many sculptors today omit the clay-modelling stage and model directly in plaster. As a mould material in the casting of concrete and fibreglass sculpture, plaster is widely used. It has great value as a material for reproducing existing

ART FOR PSYCHOTHERAPY

sculpture; many museums, for example, use such casts for study purposes.

Secondary

Basically, concrete is a mixture of an aggregate (usually sand and small pieces of stone) bound together by cement. A variety of stones, such as crushed marble, granite chips, and gravel, can be used, each giving a different effect of colour and texture.

Commercial cement is gray, white, or black; but it can be coloured by additives. The cement most widely used by sculptors is *ciment fondu*, which is extremely hard and quick setting. A recent invention—at least, in appropriate forms for sculpture—concrete is rapidly replacing stone for certain types of work.

Because it is cheap, hard, tough, and durable, it is particularly suitable for large outdoor projects, especially decorative wall surfaces. With proper reinforcement it permits great freedom of design. And by using techniques similar to those of the building industry, sculptors are able to create works in concrete on a gigantic scale.

When synthetic resins, especially polyesters, are reinforced with laminations of glass fibre, the result is a lightweight shell that is extremely strong, hard, and durable. It is usually known simply as fibreglass. After having been successfully used for car bodies, boat hulls, and the like, it has developed recently into an important material for sculpture.

Because the material is visually unattractive in itself, it is usually coloured by means of fillers and pigments. It was

ART FOR PSYCHOTHERAPY

first used in sculpture in conjunction with powdered metal fillers in order to produce cheap "cold-cast" substitutes for bronze and aluminum, but with the recent tendency to use bright colours in sculpture it is now often coloured either by pigmenting the material itself or by painting.

It is possible to model fibreglass, but more usually it is cast as a laminated shell. Its possibilities for sculpture have not yet been fully exploited.

Various formulas for modelling wax have been used in the past, but these have been generally replaced by synthetic waxes. The main uses of wax in sculpture have been as a preliminary modelling material for metal casting by the lost-wax, or cire-perdue, process and for making sketches. It is not durable enough for use as a material in its own right, although it has been used for small works, such as wax fruit, that can be kept under a glass dome.

Papier-mâché (pulped paper bonded with glue) has been used for sculpture, especially in the Far East. Mainly used for decorative work, especially masks, it can have considerable strength; the Japanese, for example, made armour from it. Sculpture made of sheet paper is a limited art form used only for ephemeral and usually trivial work.

Numerous other permanent materials—such as shells, amber, and brick—and ephemeral ones—such as feathers, baker's dough, sugar, bird seed, foliage, ice and snow, and cake icing—have been used for fashioning three-dimensional images. In view of late 20th-century trends in sculpture it is no longer possible to speak of "the materials of sculpture". Modern sculpture has no special materials.

ART FOR PSYCHOTHERAPY

Any material, natural or man-made, is likely to be used, including inflated polyethylene, foam rubber, expanded polystyrene, fabrics, and neon tubes; the materials for a sculpture by Claes Oldenburg, for example, are listed as canvas, cloth, Dacron, metal, foam rubber, and Plexiglas. Real objects, too, may be incorporated in sculpture, as in the mixed-medium compositions of Edward Kienholz; even junk has its devotees, who fashion "junk" sculpture.

Methods and techniques

Although a sculptor may specialize in, say, stone carving or direct metalwork, the art of sculpture is not identifiable with any particular craft or set of crafts. It presses into its service whatever crafts suit its purposes.

Technologies developed for more utilitarian purposes are often easily adapted for sculpture; in fact, useful artifacts and sculptured images have often been produced in the same workshop, sometimes by the same craftsman.

The methods and techniques employed in producing a pot, a bronze harness trapping, a decorative stone moulding or column, a carved wooden newel post, or even a fibreglass car body are essentially the same as those used in sculpture.

For example, the techniques of repoussé, metal casting, blacksmithing, sheet-metal work, and welding, which are used for the production of functional artifacts and decorative metalwork, are also used in metal sculpture; and the preparation, forming, glazing, decoration, and firing of clay are basically the same in both utilitarian pottery and pottery sculpture.

ART FOR PSYCHOTHERAPY

The new techniques used by sculptors today are closely related to new techniques applied in building and industrial manufacture.

Sculptor as designer and as craftsman

The conception of an artefact or a work of art—its form, imaginative content, and expressiveness—is the concern of a designer, and it should be distinguished from the execution of the work in a particular technique and material, which is the task of a craftsman. A sculptor often functions as both designer and craftsman, but these two aspects of sculpture may be separated.

Certain types of sculpture depend considerably for their aesthetic effect on the way in which their material has been directly manipulated by the artist himself. The direct, expressive handling of clay in a model by Rodin, or the use of the chisel in the stiacciato (very low) reliefs of the 15th-century Florentine sculptor Donatello could no more have been delegated to a craftsman than could the brushwork of Rembrandt.

The actual physical process of working materials is for many sculptors an integral part of the art of sculpture, and their response to the working qualities of the material—such as its plasticity, hardness, and texture—is evident in the finished work. Design and craftsmanship are intimately fused in such a work, which is a highly personal expression.

Even when the direct handling of material is not as vital as this to the expressiveness of the work, it still may be impossible to separate the roles of the artist as designer and craftsman. The qualities and interrelationships of

ART FOR PSYCHOTHERAPY

forms may be so subtle and complex that they cannot be adequately specified and communicated to a craftsman.

Moreover, many aspects of the design may actually be contributed during the process of working. Michelangelo's way of working, for example, enabled him to change his mind about important aspects of composition as the work proceeded.

A complete fusion of design and craftsmanship may not be possible if a project is a large one or if the sculptor is too old or too weak to do all of the work himself. The sheer physical labour of making a large sculpture can be considerable, and sculptors from Phidias in the 5th century BCE to Henry Moore in the 20th century, for example, have employed pupils and assistants to help with it.

Usually the sculptor delegates the time-consuming first stages of the work or some of its less important parts to his assistants and executes the final stages or the most important parts himself.

On occasion, a sculptor may function like an architect or industrial designer. He may do no direct work at all on the finished sculpture, his contribution being to supply exhaustive specifications in the form of drawings and perhaps scale models for a work that is to be entirely fabricated by craftsmen.

Obviously, such a procedure excludes the possibility of direct, personal expression through the handling of the materials; thus, works of this kind usually have the same anonymous, impersonal quality as architecture and industrial design.

ART FOR PSYCHOTHERAPY

An impersonal approach to sculpture was favoured by many sculptors of the 1960s such as William Tucker, Donald Judd, and William Turnbull. They used the skilled anonymous workmanship of industrial fabrications to make their large-scale, extremely precise, simple sculptural forms that are called "primary structures."

General methods

Broadly speaking, the stages in the production of a major work of sculpture conform to the following pattern: the commission; the preparation, submission, and acceptance of the design; the selection and preparation of materials; the forming of materials; surface finishing; installation or presentation.

Almost all of the sculpture of the past and some present-day sculpture originates in a demand made upon the sculptor from outside, usually in the form of a direct commission or through a competition. If the commission is for a portrait or a private sculpture, the client may only require to see examples of the artist's previous work; but if it is a public commission, the sculptor is usually expected to submit drawings and maquettes (small-scale, three-dimensional sketch models) that give an idea of the nature of the finished work and its relation to the site. He may be free to choose his own subject matter or theme, or it may be more or less strictly prescribed.

A medieval master sculptor, for example, received the program for a complex scheme of church sculpture from theological advisers, and Renaissance contracts for sculpture were often extremely specified and detailed. Today a great deal of sculpture is not commissioned. It arises out of the sculptor's private concern with form and imagery, and he works primarily to satisfy himself. When the work is finished he may exhibit and attempt to sell it in an art gallery.

Most of the materials used by 20th-century sculptors were readily available in a usable form from builders' or sculptors' suppliers, but certain kinds of sculpture may involve a good deal of preparatory work on the materials. A sculptor may visit a stone quarry in order to select the material for a large project and to have it cut into blocks of the right size and shape. And since stone is costly to transport and best carved when freshly quarried, he may decide to do all of his work at the quarry.

Because stone is extremely heavy, the sculptor must have the special equipment required for manoeuvring even small blocks into position for carving. A wood-carver requires a supply of well-seasoned timber and may keep a quantity of logs and blocks in store. A modeller needs a good supply of clay of the right kind. For large terra-cottas he may require a specially made-up clay body, or he may work at a brickworks, using the local clay and firing in the brick kilns.

The main part of the sculptor's work, the shaping of the material itself by modelling, carving, or constructional techniques, may be a long and arduous process, perhaps extending over a number of years and requiring assistants. Much of the work, especially architectural decoration, may be carried out at the site, or in situ.

To improve its weathering qualities, to bring out the characteristics of its material to the best advantage, or to make it more decorative or realistic, sculpture is usually given a special surface finish. It may be rubbed down and

ART FOR PSYCHOTHERAPY

polished, patinated, metal plated, gilded, painted, inlaid with other materials, and so on.

Finally, the installation of sculpture may be a complex and important part of the work. The positioning and fixing of large architectural sculpture may involve cooperation with builders and engineers; fountains may involve elaborate plumbing; the design and placing of outdoor bases, or plinths, in relation to the site and the spectator may require careful thought.

The choice of the materials, shape, and proportions of the base even for a small work requires a considerable amount of care.

Carving

Whatever material is used, the essential features of the direct method of carving are the same; the sculptor starts with a solid mass of material and reduces it systematically to the desired form. After he has blocked out the main masses and planes that define the outer limits of the forms, he works progressively over the whole sculpture, first carving the larger containing forms and planes and then the smaller ones until eventually the surface details are reached. Then he gives the surface whatever finish is required.

Even with a preliminary model as a guide, the sculptor's concept constantly evolves and clarifies as the work proceeds; thus, as he adapts his design to the nature of the carving process and the material, his work develops as an organic whole.

The process of direct carving imposes a characteristic order on the forms of sculpture. The faces of the original

ART FOR PSYCHOTHERAPY

block, slab, or cylinder of material can usually still be sensed, existing around the finished work as a kind of implied spatial envelope limiting the extension of the forms in space and connecting their highest points across space. In a similar way, throughout the whole carving, smaller forms and planes can be seen as contained within implied larger ones. Thus, an ordered sequence of containing forms and planes, from the largest to the smallest, gives unity to the work.

Indirect carving

All of the great sculptural traditions of the past used the direct method of carving, but in Western civilization during the 19th and early 20th centuries it became customary for stone and, to a lesser extent, wood sculpture to be produced by the indirect method. This required the production of a finished clay model that was subsequently cast in plaster and then reproduced in stone or wood in a more or less mechanical way by means of a pointing machine.

Usually the carving was not done by the sculptor himself. At its worst, this procedure results in a carved copy of a design that was conceived in terms of clay modelling.

Although indirect carving does not achieve aesthetic qualities that are typical of carved sculpture, it does not necessarily result in bad sculpture. Rodin's marble sculptures, for example, are generally considered great works of art even by those who object to the indirect methods by which they were produced. The indirect method has been steadily losing ground since the revival of direct carving in the early 20th century, and today it is in general disrepute among carvers.

ART FOR PSYCHOTHERAPY

Carving tools and techniques

The tools used for carving differ with the material to be carved. Stone is carved mostly with steel tools that resemble cold chisels. To knock off the corners and angles of a block, a tool called a pitcher is driven into the surface with a heavy iron hammer.

The pitcher is a thick, chisel-like tool with a wide bevelled edge that breaks rather than cuts the stone. The heavy point then does the main roughing out, followed by the fine point, which may be used to within a short distance of the final surface. These pointed tools are hammered into the surface at an angle that causes the stone to break off in chips of varying sizes.

Claw chisels, which have toothed edges, may then be worked in all directions over the surface, removing the stone in granule form and thus refining the surface forms. Flat chisels are used for finishing the surface carving and for cutting sharp detail. There are many other special tools, including stone gouges, drills, toothed hammers (known as bush hammers or bouchardes), and, often used today, power-driven pneumatic tools, for pounding away the surface of the stone. The surface can be polished with a variety of processes and materials.

Because medieval carvers worked mostly in softer stones and made great use of flat chisels, their work tends to have an edgy, cut quality and to be freely and deeply carved. In contrast is the work done in hard stones by people who lacked metal tools hard enough to cut the stone.

ART FOR PSYCHOTHERAPY

Egyptian granite sculpture, for example, was produced mainly by abrasion; that is, by pounding the surface and rubbing it down with abrasive materials. The result is a compact sculpture, not deeply hollowed out, with softened edges and flowing surfaces. It usually has a high degree of tactile appeal.

Although the process of carving is fundamentally the same for wood or stone, the physical structure of wood demands tools of a different type. For the first blocking out of a wood carving a sculptor may use saws and axes, but his principal tools are a wide range of wood-carver's gouges. The sharp, curved edge of a gouge cuts easily through the bundles of fibre and when used properly will not split the wood.

Flat chisels are also used, especially for carving sharp details. Wood rasps, or coarse files, and sandpaper can be used to give the surface a smooth finish, or, if preferred, it can be left with a faceted, chiselled appearance. Wood-carving tools have hardwood handles and are struck with round, wooden mallets. African wood sculptors use a variety of adzes rather than gouges and mallets. Ivory is carved with an assortment of saws, knives, rasps, files, chisels, drills, and scrapers.

Modelling

In contrast to the reductive process of carving, modeling is essentially a building-up process in which the sculpture grows organically from the inside. Numerous plastic materials are used for modelling. The main ones are clay, plaster, and wax; but concrete, synthetic resins, plastic wood, stucco, and even molten metal can also be modelled.

ART FOR PSYCHOTHERAPY

A design modelled in plastic materials may be intended for reproduction by casting in more permanent and rigid materials, such as metal, plaster, concrete, and fibreglass, or it may itself be made rigid and more permanent through the self-setting properties of its materials (for example, plaster) or by firing.

Modelling for casting

The material most widely used for making positive models for casting is clay. A small, compact design or a low relief can be modelled solidly in clay without any internal support; but a large clay model must be formed over a strong armature made of wood and metal. Since the armature may be very elaborate and can only be altered slightly, if at all, once work has started, the modeller must have a fairly clear idea from his drawings and maquettes of the arrangement of the main shapes of the finished model.

The underlying main masses of the sculpture are built up firmly over the armature, and then the smaller forms, surface modelling, and details are modelled over them. The modeller's chief tools are his fingers, but for fine work he may use a variety of wooden modelling tools to apply the clay and wire loop tools to cut it away.

Reliefs are modelled on a vertical or nearly vertical board. The clay is keyed, or secured, onto the board with galvanized nails or wood laths. The amount of armature required depends on the height of the relief and the weight of clay involved.

To make a cast in metal, a foundry requires from the sculptor a model made of a rigid material, usually plaster.

ART FOR PSYCHOTHERAPY

The sculptor can produce this either by modelling in clay and then casting in plaster from the clay model or by modelling directly in plaster. For direct plaster modelling, a strong armature is required because the material is brittle.

The main forms may be built up roughly over the armature in expanded wire and then covered in plaster-soaked scrim (a loosely woven sacking). This provides a hollow base for the final modelling, which is done by applying plaster with metal spatulas and by scraping and cutting down with rasps and chisels.

Fibreglass and concrete sculptures are cast in plaster moulds taken from the sculptor's original model. The model is usually clay rather than plaster because if the forms of the sculpture are at all complex it is easier to remove a plaster mould from a soft clay model than from a model in a rigid material, such as plaster.

A great deal of the metal sculpture of the past, including Nigerian, Indian, and many Renaissance bronzes, was produced by the direct lost-wax process, which involves a special modelling technique (see Casting and moulding below).

The design is first modelled in some refractory material to within a fraction of an inch of the final surface, and then the final modelling is done in a layer of wax, using the fingers and also metal tools, which can be heated to make the wax more pliable.

Medallions are often produced from wax originals, but because of their small size they can be solid-cast and therefore do not require a core.

ART FOR PSYCHOTHERAPY

Modelling for pottery sculpture

To withstand the stresses of firing, a large pottery sculpture must be hollow and of an even thickness. There are two main ways of achieving this. In the process of hollow modelling, which is typical of the potter's approach to form, the main forms of the clay model are built up directly as hollow forms with walls of a roughly even thickness.

The methods of building are similar to those employed for making hand-built pottery—coiling, pinching, and slabbing. The smaller forms and details are then added, and the finished work is allowed to dry out slowly and thoroughly before firing. The process of solid modelling is more typical of the sculptor's traditional approach to form.

The sculpture is modelled in solid clay, sometimes over a carefully considered armature, by the sculptor's usual methods of clay modelling. Then it is cut open and hollowed out, and the armature, if there is one, is removed. The pieces are then rejoined and the work is dried out and fired.

General characteristics of modelled sculpture

The process of modelling affects the design of sculpture in three important ways. First, the forms of the sculpture tend to be ordered from the inside. There are no external containing forms and planes, as in carved sculpture.

The overall design of the work—its main volumes, proportions, and axial arrangement—is determined by the underlying forms; and the smaller forms, surface

ART FOR PSYCHOTHERAPY

modelling, and decorative details are all formed around and sustained by this underlying structure.

Second, because its extension into space is not limited by the dimensions of a block of material, modelled sculpture tends to be much freer and more expansive in its spatial design than carved sculpture.

If the tensile strength of metal is to be exploited in the finished work, there is almost unlimited freedom; designs for brittle materials such as concrete or plaster are more limited.

Third, the plasticity of clay and wax encourages a fluent, immediate kind of manipulation, and many sculptors, such as Auguste Rodin, Giacomo Manzù, and Sir Jacob Epstein, like to preserve this record of their direct handling of the medium in their finished work.

Their approach contrasts with that of the Benin and Indian bronze sculptors, who refined the surfaces of their work to remove all traces of personal "handwriting."

Constructing and assembling

A constructed or assembled sculpture is made by joining preformed pieces of material. It differs radically in principle from carved and modelled sculpture, both of which are fabricated out of a homogeneous mass of material.

Constructed sculpture is made out of such basic preformed components as metal tubes, rods, plates, bars, and sheets; wooden laths, planks, dowels, and blocks; laminated timbers and chipboards; sheets of Perspex, Formica, and glass; fabrics; and wires and threads. These

ART FOR PSYCHOTHERAPY

are cut to various sizes and may be either shaped before they are assembled or used as they are.

The term *assemblage* is usually reserved for constructed sculpture that incorporates any of a vast array of ready-made, so-called found objects, such as old boilers, typewriters, engine components, mirrors, chairs, and table legs and other bits of old furniture.

Numerous techniques are employed for joining these components, most of them derived from crafts other than traditional sculptural ones; for example, metal welding and brazing, wood joinery, bolting, screwing, riveting, nailing, and bonding with new powerful adhesives.

The use of constructional techniques to produce sculpture is the main technical development of the art in recent years. Among the reasons for its popularity are that it lends itself readily to an emphasis on the spatial aspects of sculpture that preoccupied so many 20th-century artists; it is quicker than carving and modelling; it is considered by many sculptors and critics to be especially appropriate to a technological civilization; it is opening up new fields of imagery and new types of symbolism and form.

For constructed "gallery" sculpture, almost any materials and techniques are likely to be used, and the products are often extremely ephemeral. But architectural sculpture, outdoor sculpture, and indeed any sculpture that is actually used must be constructed in a safe and at least reasonably permanent manner.

The materials and techniques employed are therefore somewhat restricted. Metal sculpture constructed by riveting, bolting, and, above all, welding and brazing is best for outdoor use.

ART FOR PSYCHOTHERAPY

Direct metal sculpture

The introduction of the oxyacetylene welding torch as a sculptor's tool has revolutionized metal sculpture in recent years. A combination of welding and forging techniques was pioneered by the Spanish sculptor Julio González around 1930; and during the 1940s and 1950s it became a major sculptural technique, particularly in Britain and in the United States, where its greatest exponent was David Smith. In the 1960s and early 1970s, more sophisticated electric welding processes were replacing flame welding.

Welding equipment can be used for joining and cutting metal. A welded joint is made by melting and fusing together the surfaces of two pieces of metal, usually with the addition of a small quantity of the same metal as filler.

The metal most widely used for welded sculpture is mild steel, but other metals can be welded. In a brazed joint, the parent metals are not actually fused together but are joined by an alloy that melts at a lower temperature than the parent metals.

Brazing is particularly useful for making joints between different kinds of metal, which cannot be done by welding, and for joining nonferrous metals. Forging is the direct shaping of metal by bending, hammering, and cutting.

Direct metalworking techniques have opened up whole new ranges of form to the sculptor—open skeletal structures, linear and highly extended forms, and complex, curved sheet forms. Constructed metal sculpture may be precise and clean, as that of Minimalist

ART FOR PSYCHOTHERAPY

sculptors Donald Judd and Phillip King, or it may exploit the textural effects of molten metal in a free, "romantic" manner.

Reproduction and surface-finishing techniques

Casting and moulding processes are used in sculpture either for making copies of existing sculpture or as essential stages in the production of a finished work. Numerous materials are used for making moulds and casts, and some of the methods are complex and highly skilled. Only a broad outline of the principal methods can be given here.

Casting and moulding

These are used for producing a single cast from a soft, plastic original, usually clay. They are especially useful for producing master casts for subsequent reproduction in metal. The basic procedure is as follows.

First, the mould is built up in liquid plaster over the original clay model; for casting reliefs, a one-piece mould may be sufficient, but for sculpture in the round a mould in at least two sections is required.

Second, when the plaster is set, the mold is divided and removed from the clay model.

Third, the mould is cleaned, reassembled, and filled with a self-setting material such as plaster, concrete, or fibreglass-reinforced resin.

Fourth, the mould is carefully chipped away from the cast.

ART FOR PSYCHOTHERAPY

This involves the destruction of the mould—hence the term "waste" mould. The order of reassembling and filling the mould may be reversed; fibreglass and resin, for example, are "laid up" in the mould pieces before they are reassembled.

Plaster piece moulds are used for producing more than one cast from a soft or rigid original and are especially good for reproducing existing sculpture and for slip casting (see below). Before the invention of flexible moulds (see below), piece moulds were used for producing wax casts for metal casting by the lost-wax process.

A piece mould is built up in sections that can be withdrawn from the original model without damaging it. The number of sections depends on the complexity of the form and on the amount of undercutting; tens, or even hundreds, of pieces may be required for really large, complex works.

The mould sections are carefully keyed together and supported by a plaster case. When the mould has been filled, it can be removed section by section from the cast and used again. Piece moulding is a highly skilled and laborious process.

Made of such materials as gelatine, vinyl, and rubber, flexible moulds are used for producing more than one cast; they offer a much simpler alternative to piece moulding when the original model is a rigid one with complex forms and undercuts. The material is melted and poured around the original positive in sections, if necessary.

Being flexible, the mould easily pulls away from a rigid surface without causing damage. While it is being filled

ART FOR PSYCHOTHERAPY

(with wax, plaster, concrete, and fibreglass-reinforced resins), the mould must be surrounded by a plaster case to prevent distortion.

The lost-wax process is the traditional method of casting metal sculpture. It requires a positive, which consists of a core made of a refractory material and an outer layer of wax. The positive can be produced either by direct modelling in wax over a prepared core, in which case the process is known as direct lost-wax casting, or by casting in a piece mould or flexible mould taken from a master cast.

The wax positive is invested with a mould made of refractory materials and is then heated to a temperature that will drive off all moisture and melt out all the wax, leaving a narrow cavity between the core and the investment. Molten metal is then poured into this cavity. When the metal has cooled down and solidified, the investment is broken away, and the core is removed from inside the cast.

The process is, of course, much more complex than this simple outline suggests. Care has to be taken to suspend the core within the mould by means of metal pins, and a structure of channels must be made in the mould that will enable the metal to reach all parts of the cavity and permit the mould gases to escape. A considerable amount of filing and chasing of the cast is usually required after casting is completed.

While the lost-wax process is used for producing complex, refined metal castings, sand moulding is more suitable for simpler types of form and for sculpture in which a certain roughness of surface does not matter.

ART FOR PSYCHOTHERAPY

Recent improvements in the quality of sand castings and the invention of the "lost-pattern" process have resulted in a much wider use of sand casting as a means of producing sculpture.

A sand mould, made of special sand held together by a binder, is built up around a rigid positive, usually in a number of sections held together in metal boxes.

For a hollow casting, a core is required that will fit inside the negative mould, leaving a narrow cavity as in the lost-wax process. The molten metal is poured into this cavity.

The lost-pattern process is used for the production by sand moulding of single casts in metal. After a positive made of expanded polystyrene is firmly embedded in casting sand, molten metal is poured into the mould straight onto the expanded foam original.

The heat of the metal causes the foam to pass off into vapour and disappear, leaving a negative mould to be filled by the metal. Channels for the metal to run in and for the gases to escape are made in the mould, as in the lost-wax process. The method is used mainly for producing solid castings in aluminum that can be welded or riveted together to make the finished sculpture.

Slip casting is primarily a potter's technique that can be used for repetition casting of small pottery sculptures. Liquid clay, or slip, is poured into a plaster piece mould.

Some of the water in the slip is absorbed by the plaster and a layer of stiffened clay collects on the surface of the mould. When this layer is thick enough to form a cast, the excess slip is poured off and the mould is removed. The hollow clay cast is then dried and fired.

ART FOR PSYCHOTHERAPY

Simple casts for pottery sculpture—mainly tiles and low reliefs—can be prepared by pressing clay into a rigid mould. More complex forms can be built up from a number of separately press-cast pieces. Simple terra-cotta moulds can be made by pressing clay around a rigid positive form. After firing, these press molds can be used for press casting.

Pointing

A sculpture can be reproduced by transposing measurements taken all over its surface to a copy. The process is made accurate and thorough by the use of a pointing machine, which is an arrangement of adjustable metal arms and pointers that are set to the position of any point on the surface of a three-dimensional form and then used to locate the corresponding point on the surface of a copy.

If the copy is a stone one, the block is drilled to the depth measured by the pointing machine. When a number of points have been fixed by drilling, the stone is cut away to the required depth. For accurate pointing, a vast number of points have to be taken, and the final surface is approached gradually. The main use of pointing has been for the indirect method of carving.

Enlarged and reduced copies of sculpture can also be produced with the aid of mechanical devices. A sophisticated reducing machine that works on the principle of the pantograph (an instrument for copying on any predetermined scale, consisting of four light, rigid bars jointed in parallelogram form) is used in minting for scaling down the sculptor's original model to coin size.

ART FOR PSYCHOTHERAPY

Surface finishing

Surface finishes for sculpture can be either natural—bringing the material of the sculpture itself to a finish—or applied. Almost all applied surface finishes preserve as well as decorate.

Smoothing and polishing

Many sculptural materials have a natural beauty of colour and texture that can be brought out by smoothing and polishing. Stone carvings are smoothed by rubbing down with a graded series of coarse and fine abrasives, such as carborundum, sandstone, emery, pumice, and whiting, all used while the stone is wet.

Some stones, such as marble and granite, will take a high gloss; others are too coarse-grained to be polished and can only be smoothed to a granular finish. Wax is sometimes used to give stone a final polish.

The natural beauty of wood is brought out by sandpapering or scraping and then waxing or oiling. Beeswax and linseed oil are the traditional materials, but a wide range of waxes and oils is currently available.

Ivory is polished with gentle abrasives such as pumice and whiting, applied with a damp cloth.

Concrete can be rubbed down, like stone, with water and abrasives, which both smooth the surface and expose the aggregate. Some concretes can be polished.

Metals are rubbed down manually with steel wool and emery paper and polished with various metal polishes. A high-gloss polish can be given to metals by means of

ART FOR PSYCHOTHERAPY

power-driven buffing wheels used in conjunction with abrasives and polishes. Clear lacquers are applied to preserve the polish.

Painting

Stone, wood, terra-cotta, metal, fibreglass, and plaster can all be painted in a reasonably durable manner provided that the surfaces are properly prepared and suitable primings and paints are used. In the past, stone and wood carvings were often finished with a coating of gesso (plaster of Paris or gypsum prepared with glue) that served both as a final modelling material for delicate surface detail and as a priming for painting.

Historically, the painting and gilding of sculpture were usually left to specialists. In Greek relief sculpture, actual details of the composition were often omitted at the carving stage and left for the painter to insert. In the 15th century, the great Flemish painter Rogier van der Weyden undertook the painting of sculpture as part of his work.

Modern paint technology has made an enormous range of materials available. Constructed sculptures are often finished with mechanical grinders and sanders and then sprayed with high-quality cellulose paints.

Gilding

The surfaces of wood, stone, and plaster sculpture can be decorated with gold, silver, and other metals that are applied in leaf or powder form over a suitable priming.

ART FOR PSYCHOTHERAPY

Metals, especially bronze, were often fire-gilded; that is, treated with an amalgam of gold and mercury that was heated to drive off the mercury. The panels of the *Gates of Paradise* in Florence, by the 15th-century sculptor Lorenzo Ghiberti, are a well-known example of gilded bronze.

Patination

Patinas on metals are caused by the corrosive action of chemicals. Sculpture that is exposed to different kinds of atmosphere or buried in soil or immersed in seawater for some time acquires a patina that can be extremely attractive. Similar effects can be achieved artificially by applying various chemicals to the metal surface, which is often heated to create a bond.

This is a particularly effective treatment for bronze, which can be given a wide variety of attractive green, brown, blue, and black patinas. Iron is sometimes allowed to rust until it acquires a satisfactory colour, and then the process is arrested by lacquering.

Electroplating

The surfaces of metal sculpture or of specially prepared non-metal sculpture can be coated with such metals as chrome, silver, gold, copper, and nickel by the familiar industrial process of electroplating. The related technique of anodizing can be used to prevent the corrosion of aluminum sculpture and to dye its surface.

ART FOR PSYCHOTHERAPY

Other finishes

The surfaces of metal sculpture can be decorated by means of numerous metal-smithing techniques—etching, engraving, metal inlaying, enamelling, and so on. Pottery sculpture can be decorated with coloured slips, oxides, and enamels; glazed with a variety of shiny or mat glazes; and brought to a dull polish by burnishing.

Other materials have often been added to the surface of sculpture. The eyes of ancient figure sculpture, for example, were sometimes inlaid with stones. Occasionally—as in Mexican mosaic work—the whole surface of a sculpture is inlaid with mother-of-pearl, turquoise, coral, and many other substances.

Symbolism of sculpture

A great deal of sculpture is designed to be placed in public squares, gardens, parks, and similar open places or in interior positions where it is isolated in space and can be viewed from all directions. Other sculpture is carved in relief and is viewed only from the front and sides.

Sculpture in the round

The opportunities for free spatial design that such freestanding sculpture presents are not always fully exploited. The work may be designed, like many Archaic sculptures, to be viewed from only one or two fixed positions, or it may in effect be little more than a four-sided relief that hardly changes the three-dimensional form of the block at all.

Sixteenth-century Mannerist sculptors, on the other hand, made a special point of exploiting the all-around visibility of freestanding sculpture. Giambologna's *Rape of the Sabines*, for example, compels the viewer to walk all around it in order to grasp its spatial design.

It has no principal views; its forms move around the central axis of the composition, and their serpentine movement unfolds itself gradually as the spectator moves around to follow them.

Much of the sculpture of Henry Moore and other 20th-century sculptors is not concerned with movement of this kind, nor is it designed to be viewed from any fixed positions.

Rather, it is a freely designed structure of multidirectional forms that is opened up, pierced, and extended in space in such a way that the viewer is made aware of its all-around design largely by seeing through the sculpture.

The majority of constructed sculptures are disposed in space with complete freedom and invite viewing from all directions. In many instances the spectator can actually walk under and through them.

The way in which a freestanding sculpture makes contact with the ground or with its base is a matter of considerable importance. A reclining figure, for example, may in effect be a horizontal relief. It may blend with the ground plane and appear to be rooted in the ground like an outcrop of rock.

Other sculptures, including some reclining figures, may be designed in such a way that they seem to rest on the

ART FOR PSYCHOTHERAPY

ground and to be independent of their base. Others are supported in space above the ground. The most completely freestanding sculptures are those that have no base and may be picked up, turned in the hands, and literally viewed all around like a netsuke (a small toggle of wood, ivory, or metal used to fasten a small pouch or purse to a kimono sash).

Of course, a large sculpture cannot actually be picked up in this way, but it can be designed so as to invite the viewer to think of it as a detached, independent object that has no fixed base and is designed all around.

Sculpture designed to stand against a wall or similar background or in a niche may be in the round and freestanding in the sense that it is not attached to its background like a relief; but it does not have the spatial independence of completely freestanding sculpture, and it is not designed to be viewed all around.

It must be designed so that its formal structure and the nature and meaning of its subject matter can be clearly apprehended from a limited range of frontal views. The forms of the sculpture, therefore, are usually spread out mainly in a lateral direction rather than in depth. Greek pedimental sculpture illustrates this approach superbly: the composition is spread out in a plane perpendicular to the viewer's line of sight and is made completely intelligible from the front. Seventeenth-century Baroque sculptors, especially Bernini, adopted a rather different approach.

Though some favoured a coherent frontal viewpoint, however active, Bernini is known to have conceived a work (the *Apollo and Daphne*) in which the narrative

ART FOR PSYCHOTHERAPY

unfolded in details discovered as the viewer walked around the work, beginning from the rear.

The frontal composition of wall and niche sculpture does not necessarily imply any lack of three-dimensionality in the forms themselves; it is only the arrangement of the forms that is limited.

Classical pedimental sculpture, Indian temple sculpture such as that at Khajuraho, Gothic niche sculpture, and Michelangelo's Medici tomb figures are all designed to be placed against a background, but their forms are conceived with a complete fullness of volume.

Relief sculpture

Relief sculpture is a complex art form that combines many features of the two-dimensional pictorial arts and the three-dimensional sculptural arts. On the one hand, a relief, like a picture, is dependent on a supporting surface, and its composition must be extended in a plane in order to be visible. On the other hand, its three-dimensional properties are not merely represented pictorially but are in some degree actual, like those of fully developed sculpture.

Among the various types of relief are some that approach very closely the condition of the pictorial arts. The reliefs of Donatello, Ghiberti, and other early Renaissance artists make full use of perspective, which is a pictorial method of representing three-dimensional spatial relationships realistically on a two-dimensional surface.

Egyptian and most pre-Columbian American low reliefs are also extremely pictorial but in a different way. Using a system of graphic conventions, they translate the three-

ART FOR PSYCHOTHERAPY

dimensional world into a two-dimensional one. The relief image is essentially one of plane surfaces and could not possibly exist in three dimensions. Its only sculptural aspects are its slight degree of actual projection from a surface and its frequently subtle surface modelling.

Other types of relief—for example, Classical Greek and most Indian—are conceived primarily in sculptural terms. The figures inhabit a space that is defined by the solid forms of the figures themselves and is limited by the background plane.

This back plane is treated as a finite, impenetrable barrier in front of which the figures exist. It is not conceived as a receding perspective space or environment within which the figures are placed nor as a flat surface upon which they are placed. The reliefs, so to speak, are more like contracted sculpture than expanded pictures.

The central problem of relief sculpture is to contract or condense three-dimensional solid form and spatial relations into a limited depth space. The extent to which the forms actually project varies considerably, and reliefs are classified on this basis as low reliefs (bas-reliefs) or high reliefs.

There are types of reliefs that form a continuous series from the almost completely pictorial to the almost fully in the round.

One of the relief sculptor's most difficult tasks is to represent the relations between forms in depth within the limited space available to him. He does this mainly by giving careful attention to the planes of the relief.

In a carved relief the highest, or front, plane is defined by the surface of the slab of wood or stone in which the

ART FOR PSYCHOTHERAPY

relief is carved; and the back plane is the surface from which the forms project.

The space between these two planes can be thought of as divided into a series of planes, one behind the other. The relations of forms in depth can then be thought of as relations between forms lying in different planes.

Sunken relief is also known as incised, coelanaglyphic, and intaglio relief. It is almost exclusively an ancient Egyptian art form, but some beautiful small-scale Indian examples in ivory have been discovered at Bagrām in Afghanistan. In a sunken relief, the outline of the design is first incised all around. The relief is then carved inside the incised outline, leaving the surrounding surface untouched.

Thus, the finished relief is sunk below the level of the surrounding surface and is contained within a sharp, vertical-walled contour line. This approach to relief sculpture preserves the continuity of the material's original surface and creates no projection from it. The outline shows up as a powerful line of light and shade around the whole design.

Modern forms of sculpture

Since the 1950s, many new combined forms of art have been developed that do not fit readily into any of the traditional categories. Two of the most important of these, environments and kinetics, are closely enough connected with sculpture to be regarded by many artists and critics as branches or offshoots of sculpture.

It is likely, however, that the persistence of the terms environmental sculpture and kinetic sculpture is a result

ART FOR PSYCHOTHERAPY

of the failure of language to keep pace with events; for the practice is already growing of referring simply to environments and kinetics, as one might refer to painting, sculpture, and engraving, as art forms in their own right.

Traditional sculptures in relief and in the round are static, fixed objects or images. Their immobility and immutability are part of the permanence traditionally associated with the art of sculpture, especially monumental sculpture. What one refers to as movement in, say, a Baroque or Greek sculpture is not actual physical motion but a movement that is either directly represented in the subject matter (galloping horses) or expressed through the dynamic character of its form (spirals, undulating curves).

In recent years, however, the use of actual movement, kineticism, has become an important aspect of sculpture. Naum Gabo, Marcel Duchamp, László Moholy-Nagy, and Alexander Calder were pioneers of kinetic sculpture in modern times, but many kinetic artists see a connection between their work and such forms as the moving toys, dolls, and clocks of previous ages.

There are now types of sculpture in which the components are moved by air currents, as in the well-known mobiles of Calder; by water; by magnetism, the speciality of Nicholus Takis; by a variety of electromechanical devices; or by the participation of the spectator himself.

The neo-Dada satire quality of the kinetic sculpture created during the 1960s is exemplified by the works of Jean Tinguely. His self-destructing "Homage to New York" perfected the concept of a sculpture being both an object and an event, or "happening".

ART FOR PSYCHOTHERAPY

The aim of most kinetic sculptors is to make movement itself an integral part of the design of the sculpture and not merely to suggest movement within a static object. Calder's mobiles, for example, depend for their aesthetic effect on constantly changing patterns of relationship.

When liquids and gases are used as components, the shapes and dimensions of the sculpture may undergo continual transformations.

The movement of smoke; the diffusion and flow of coloured water, mercury, oil, and so on; pneumatic inflation and deflation; and the movement of masses of bubbles have all served as media for kinetic sculpture. In the complex, electronically controlled "spatio-dynamic" and "lumino-dynamic" constructions of Nicolas Schöffer, the projection of changing patterns of light into space is a major feature.

The environmental sculptor creates new spatial contexts that differ from anything developed by traditional sculpture. The work no longer confronts the spectator as an object but surrounds him so that he moves within it as he might within a stage set, a garden, or an interior.

The most common type of environment is the "room", which may have specially shaped and surfaced walls, special lighting effects, and many different kinds of contents.

Kurt Schwitters's *Merzbau* (destroyed in 1943) was the first of these rooms, which now include the nightmare fantasy of Edward Kienholz's tableaux, such as *Roxy's* (1961) or *The Illegal Operation* (1962); George Segal's compositions, in which casts of clothed human figures in frozen, casual attitudes are placed in interiors; and rooms built of mirrors, such as Yayoi Kusama's *Endless Love*

ART FOR PSYCHOTHERAPY

Room and Lucas Samaras's *Mirrored Room*, in both of which the spectator himself, endlessly reflected, becomes part of the total effect.

Environmental art, in common with collage and assemblage, has tended toward greater concreteness not by making a more realistic representation, as naturalistic art does, but by including more of reality itself in the work; for example, by using casts taken from the actual human body, real clothes, actual objects and casts of objects, actual lighting effects, and real items of furniture.

Plastic elements may be combined with music and sound effects, dance, theatrical spectacles, and film to create so-called happenings, in which real figures are constituents of the "artwork" and operations are performed not on "artistic" materials but are performed on real objects and on the actual environment. Ideas such as these go far beyond anything that has ever before been associated with the term sculpture.

Representational sculpture

Sculpture in the round is much more restricted than relief in the range of its subject matter. The representation of, say, a battle scene or a cavalcade in the round would require a space that corresponded in scale in every direction with that occupied by an actual battle or cavalcade.

No such problems arise in relief because the treatment of scale and relations in depth is to some extent notional, or theoretical, like that of pictures. Then again, because a relief is attached to a background, problems of weight and physical balance and support do not arise. Figures

ART FOR PSYCHOTHERAPY

can be represented as floating in space and can be arranged vertically as well as horizontally.

Thus, in general, sculpture in the round is concerned with single figures and limited groups, while reliefs deal with more complex "pictorial" subjects involving crowds, landscape, architectural backgrounds, and so on.

Human figure

The principal subject of sculpture has always been the human figure. Next in importance in historical work are animals and fantastic creatures based on human and animal forms. Other subjects—for example, landscape, plants, still life, and architecture—have served primarily as accessories to figure sculpture, not as subjects in their own right, except as decorative elements within architecture or as precious carved witticisms such as those of the British wood-carver Grinling Gibbons.

The overwhelming predominance of the human figure is due: first, to its immense emotional importance as an object of desire, love, fear, respect, and, in the case of anthropomorphic gods, worship; and, second, to its inexhaustible subtlety and variety of form and expression.

The nude or almost nude figure played a prominent role in Egyptian, Indian, Greek, and African sculpture, while in medieval European and ancient Chinese sculpture the figure is almost invariably clothed. The interplay of the linear and modelled forms of free draperies with the solid volumes of the human body was of great interest to Classical sculptors and later became one of the principal themes of Renaissance and post-Renaissance sculpture.

ART FOR PSYCHOTHERAPY

The human figure continues to be of central importance in modern sculpture in spite of the growth of non-figurative art; but the optimistic, idealized, or naturalistic images of man prevalent in previous ages have been largely replaced by images of despair, horror, deformation, and satire.

Devotional images and narrative sculpture

The production of devotional images has been one of the sculptor's main tasks, and many of the world's greatest sculptures are of this kind. They include images of Buddha and the Hindu gods; of Christ, the Virgin, and the Christian saints; of Athena, Aphrodite, Zeus, and other Greek gods; and of all the various gods, spirits, and mythical beings of Rome, the ancient Near East, pre-Columbian America, Black Africa, and the Pacific Islands.

Closely connected with devotional images are all of the commemorative narrative sculptures in which legends, heroic deeds, and religious stories are depicted for the delight and instruction of peoples who lived when books and literacy were rare.

The Buddhist, Hindu, and Christian traditions are especially rich in narrative sculpture. Stories of the incarnations of Buddha—*Jataka*—and of the Hindu gods abounded in the temple sculpture of India and Southeast Asia; for example, at Sanchi, Amaravati, Borobudur, and Angkor. Sculpture illustrating the stories of the Bible is so abundant in medieval churches that the churches have been called "Bibles in stone." Sculpture recounting the heroic deeds of kings and generals are common, especially in Assyria and Rome.

ART FOR PSYCHOTHERAPY

The Romans made use of a form known as continuous narrative, the best known example of which is the spiral, or helical, band of relief sculpture that surrounds Trajan's Column (*c.* AD 106–113) and tells the story of the Emperor's Dacian Wars. The episodes in the narrative are not separated into a series of framed compositions but are linked to form a continuous band of unbroken relief.

Portraiture

Portraiture was practiced by the Egyptians but was comparatively rare in the ancient world until the Greeks and Romans made portrait sculpture one of their major artistic achievements.

The features of many famous people are known to modern man only through the work of Roman sculptors on coins and medals, portrait busts, and full-length portraits. Portraiture has been an important aspect of Western sculpture from the Renaissance to the present day. Some of the best known modern portrait sculptors are Rodin, Charles Despiau, Marino Marini, and Jacob Epstein.

Scenes of everyday life

Scenes of everyday life have been represented in sculpture mainly on a small scale in minor works.

The sculptures that are closest in spirit to the quiet dignity of the great 17th- and 18th-century genre paintings of Johannes Vermeer and Jean-Baptiste-Siméon Chardin are perhaps certain Greek tombstones, such as that of the Stele of Hegeso, which represents a quiet, absorbed moment when a seated young woman

ART FOR PSYCHOTHERAPY

and her maidservant are looking at a necklace they have just removed from a casket. Intimate scenes of the people and their activities in everyday rural life are often portrayed in medieval and Egyptian reliefs as part of larger compositions.

Animals

Animals have always been important subjects for sculpture. Palaeolithic man produced some extraordinarily sensitive animal sculptures both in relief and in the round. Representations of horses and lions are among the finest works of Assyrian sculpture.

Egyptian sculptors produced sensitive naturalistic representations of cattle, donkeys, hippopotamuses, apes, and a wide variety of birds and fish. Ancient Chinese sculptors made superb small-scale animal sculptures in bronze and pottery.

Animals were the main subject matter for the sculpture of the nomadic tribes of Eurasia and northern Europe, for whom they became the basis for elaborate zoomorphic fantasies. This animal art contributed to the rich tradition of animal sculpture in medieval art. Animals also served as a basis for semi-abstract fantasy in Mexican, Maya, North American Indian, and Oceanic sculpture.

The horse has always occupied an important place in Western sculpture, but other animals have also figured in the work of such sculptors as Giambologna, in the 16th century, and Antoine-Louis Barye, in the 19th, as well as numerous sculptors of garden and fountain pieces.

Among modern sculptors who have made extensive use of animals or animal-like forms are Brancusi, Picasso,

ART FOR PSYCHOTHERAPY

Gerhard Marcks, Germaine Richier, François Pompon, Pino Pascali, and François-Xavier-Maxime Lalanne.

Fantasy

In their attempts to imagine gods and mythical beings, sculptors have invented fantastic images based on the combination and metamorphosis of animal and human forms. A centaur, the Minotaur, and animal-headed gods of the ancient world are straightforward combinations.

More imaginative fantasies were produced by Mexican and Maya sculptors and by tribal sculptors in many parts of the world. Fantastic creatures abound in the sculpture produced in northern Europe during the early Middle Ages and the Romanesque period. Fantasy of a playful kind is often found in garden sculpture and fountains.

In the period following World War I, fantasy was a dominant element in representational sculpture. Among its many forms are images derived from dreams, the technological fantasy of science fiction, erotic fantasies, and a whole host of monsters and automata. The Surrealists have made a major contribution to this aspect of modern sculpture.

Other subjects

Architectural backgrounds in sculpture range from the simplified baldacchinos (ornamental structures resembling canopies used especially over altars) of early medieval reliefs to the 17th- and 18th-century virtuoso perspective townscapes of Grinling Gibbons.

ART FOR PSYCHOTHERAPY

Architectural accessories such as plinths, entablatures, pilasters, columns, and mouldings have played a prominent role both in Greek and Roman sarcophagi, in medieval altarpieces and screens, and in Renaissance wall tombs.

Outside the field of ornament, botanical forms have played only a minor role in sculpture. Trees and stylized lotuses are especially common in Indian sculpture because of their great symbolic significance. Trees are also present in many Renaissance reliefs and in some medieval reliefs.

Landscape, which was an important background feature in many Renaissance reliefs (notably those of Ghiberti) and, as sculptured rocks, appeared in a number of Baroque fountains, entered into sculpture in a new way when Henry Moore combined the forms of caves, rocks, hills, and cliffs with the human form in a series of large reclining figures.

There is nothing in sculpture comparable with the tradition of still-life painting. When objects are represented, it is almost always as part of a figure composition. A few modern sculptors, however, notably Giacomo Manzù and Oldenburg, have used still-life subjects.

Non-representational sculpture

There are two main kinds of non-representational sculpture. One kind uses nature not as subject matter to be represented but as a source of formal ideas. For sculptors who work in this way, the forms that are observed in nature serve as a starting point for a kind of

ART FOR PSYCHOTHERAPY

creative play, the end products of which may bear little or no resemblance either to their original source or to any other natural object. Many works by Brancusi, Raymond Duchamp-Villon, Jacques Lipchitz, Henri Laurens, Umberto Boccioni, and other pioneer modern sculptors have this character.

The transformation of natural forms to a point where they are no longer recognizable is also common in many styles of primitive and ornamental art.

The other main kind of non-representational sculpture, often known as non-objective sculpture, is a more completely non-representational form that does not even have a starting point in nature. It arises from a constructive manipulation of the sculptor's generalized, abstract ideas of spatial relations, volume, line, colour, texture, and so on.

The approach of the non-objective sculptor has been likened to that of the composer of music, who manipulates the elements of his art in a similar manner. The inclusion of purely invented, three-dimensional artifacts under the heading of sculpture is a 20th-century innovation.

Some non-objective sculptors prefer forms that have the complex curvilinearity of surface typical of living organisms; others prefer more regular, simple geometric forms. The whole realm of three-dimensional form is open to non-objective sculptors, but these sculptors often restrict themselves to a narrow range of preferred types of form.

A kind of non-objective sculpture prominent in the 1950s and '60s, for example, consisted of extremely stark, so-called primary forms. These were highly finished, usually coloured constructions that were often large in scale and made up entirely of plane or single-curved surfaces.

Prominent among the first generation of non-objective sculptors were Jean Arp, Antoine Pevsner, Naum Gabo, Barbara Hepworth, Max Bill, and David Smith. Subsequent artists who worked in this manner include Robert Morris, Donald Judd, and Phillip King.

Decorative sculpture

The devices and motifs of ornamental sculpture fall into three main categories: abstract, zoomorphic, and botanical. Abstract shapes, which can easily be made to fit into any framework, are a widespread form of decoration.

Outstanding examples of abstract relief ornament are found on Islamic, Mexican, and Maya buildings and on small Celtic metal artefacts. The character of the work varies from the large-scale rectilinear two-plane reliefs of the buildings of Mitla in Mexico, to the small-scale curvilinear plastic decoration of a Celtic shield or body ornament.

Zoomorphic relief decoration, derived from a vast range of animal forms, is common on primitive artefacts and on Romanesque churches, especially the wooden stave churches of Scandinavia.

Botanical forms lend themselves readily to decorative purposes because their growth patterns are variable and their components—leaf, tendril, bud, flower, and fruit—are infinitely repeatable.

The acanthus and anthemion motifs of Classical relief and the lotuses of Indian relief are splendid examples of

ART FOR PSYCHOTHERAPY

stylized plant ornament. The naturalistic leaf ornament of Southwell Minster, Reims Cathedral, and other Gothic churches transcends the merely decorative and becomes superbly plastic sculpture in its own right.

Symbolism

Sculptural images may be symbolic on a number of levels. Apart from conventional symbols, such as those of heraldry and other insignia, the simplest and most straightforward kind of sculptural symbol is that in which an abstract idea is represented by means of allegory and personification.

A few common examples are figures that personify the cardinal virtues (prudence, justice, temperance, fortitude), the theological virtues (faith, hope, and charity), the arts, the church, victory, the seasons of the year, industry, and agriculture.

These figures are often provided with symbolic objects that serve to identify them; for example, the hammer of industry, the sickle of agriculture, the hourglass of time, and the scales of justice. Such personifications abound in medieval and Renaissance sculpture and were until recently the stock in trade of public sculpture the world over. Animals are also frequently used in the same way; for example, the owl (as the emblem of Athens and the symbol of wisdom), the British lion, and the American eagle.

Beyond this straightforward level of symbolism, the images of sculpture may serve as broader, more abstruse religious, mythical, and civic symbols expressing some

ART FOR PSYCHOTHERAPY

of mankind's deepest spiritual insights, beliefs, and feelings.

The great tympanums (the space above the lintel of a door that is enclosed by the doorway arch) of Autun, Moissac, and other medieval churches symbolize some of the most profound Christian doctrines concerning the ends of human life and man's relations with the divine.

The Hindu image of the dance of Shiva is symbolic in every detail, and the whole image expresses in one concentrated symbol some of the complex cosmological ideas of the Hindu religion.

The Buddhist temple of Borobudur, in Java, is one of the most complex and integrated of all religious symbols. It is designed as a holy mountain whose structure symbolizes the structure of the spiritual universe.

Each of the nine levels of the temple has a different kind of sculptural symbolism, progressing from symbols of hell and the world of desire at the lowest level to austere symbols of the higher spiritual mysteries at its uppermost levels.

In more individualistic societies, works of sculpture may be symbolic on a personal, private level. Michelangelo's "Slaves" have been interpreted as Neo-platonic allegories of the human soul struggling to free itself from the bondage of the body, its "earthly prison," or, more directly, as symbols of the struggle of intelligible form against mere matter.

But there is no doubt that, in ways difficult to formulate precisely, they are also disturbing symbols of Michelangelo's personal attitudes, emotions, and psychological conflicts. If it is an expression of his

ART FOR PSYCHOTHERAPY

unconscious mind, the sculptor himself may be unaware of this aspect of the design of his work.

Many modern sculptors disclaim any attempt at symbolism in their work. When symbolic images do play a part in modern sculpture, they are either derived from obsolete classical, medieval, and other historical sources or they are private.

Because there has been little socially recognized symbolism for the modern sculptor to use in his work, symbols consciously invented by individual artists or deriving from the image-producing function of the individual unconscious mind have been paramount.

Many of these are entirely personal symbols expressing the artist's private attitudes, beliefs, obsessions, and emotions. They are often more symptomatic than symbolic.

Henry Moore is outstanding among modern sculptors for having created a world of personal symbols that also have a universal quality; and Naum Gabo has sought images that would symbolize in a general way modern man's attitudes to the world picture provided by science and technology.

Examples of sculpture of which the positioning, or siting, as well as the imagery is symbolic are the carved boundary stones of the ancient world; memorials sited on battlegrounds or at places where religious and political martyrs have been killed; the Statue of Liberty and similar civic symbols situated at harbours, town gates, bridges, and so on; and the scenes of the Last Judgment placed over the entrances to cathedrals, where they could serve as an admonition to the congregation.

ART FOR PSYCHOTHERAPY

The choice of symbolism suitable to the function of a sculpture is an important aspect of design. Fonts, pulpits, lecterns, triumphal arches, war memorials, tombstones, and the like all require a symbolism appropriate to their function. In a somewhat different way, the tomb sculptures of Egypt, intended to serve a magical function in the afterlife of the tomb's inhabitants, had to be images suitable for their purpose. These, however, are more in the nature of magical substitutes than symbols.

Uses of sculpture

The vast majority of sculptures are not entirely autonomous but are integrated or linked in some way with other works of art in other mediums. Relief, in particular, has served as a form of decoration for an immense range of domestic, personal, civic, and sacred artefacts, from the spear-throwers of Palaeolithic man and the cosmetic palettes of earliest Egyptian civilization to the latest mass-produced plastic reproduction of a Jacobean linen-fold panel (a carved or moulded panel representing a fold, or scroll, of linen).

The main use of large-scale sculpture has been in conjunction with architecture. It has either formed part of the interior or exterior fabric of the building itself or has been placed against or near the building as an adjunct to it.

The role of sculpture in relation to buildings as part of a townscape is also of considerable importance. Traditionally, it has been used to provide a focal point at the meeting of streets, and in marketplaces, town squares, and other open places—a tradition that many town planners today are continuing.

ART FOR PSYCHOTHERAPY

Sculpture has been widely used as part of the total decorative scheme for a garden or park. Garden sculpture is usually intended primarily for enjoyment, helping to create the right kind of environment for meditation, relaxation, and delight.

Because the aim is to create a light-hearted Arcadian or ideal paradisal atmosphere, disturbing or serious subjects are usually avoided. The sculpture may be set among trees and foliage where it can surprise and delight the viewer or sited in the open to provide a focal point for a vista.

Fountains, too, are intended primarily to give enjoyment to the senses. There is nothing to compare with the interplay of light, movement, sound, and sculptural imagery in great fountains, which combine the movement and sound of sheets, jets, and cataracts of water with richly imaginative sculpture, water plants and foliage, darting fish, reflections, and changing lights. They are the prototypes of all 20th-century "mixed-media" kinetic sculptures.

The durability of sculpture makes it an ideal medium for commemorative purposes, and much of the world's greatest sculpture has been created to perpetuate the memory of persons and events. Commemorative sculpture includes tombs, tombstones, statues, plaques, sarcophagi, memorial columns, and triumphal arches. Portraiture, too, often serves a memorial function.

One of the most familiar and widespread uses of sculpture is for coins. Produced for more than 2,500 years, these miniature works of art contain a historically invaluable and often artistically excellent range of portrait heads and symbolic devices.

ART FOR PSYCHOTHERAPY

Medals, too, in spite of their small scale, may be vehicles for plastic art of the highest quality. The 15th-century medals of the Italian artist Antonio Pisanello and the coins of ancient Greece are generally considered the supreme achievements in these miniature fields of sculpture.

Also on a small scale are the sculptural products of the glyptic arts—that is, the arts of carving gems and hard stones. Superb and varied work, often done in conjunction with precious metalwork, has been produced in many countries.

Finally, sculpture has been widely used for ceremonial and ritualistic objects such as bishop's crosiers, censers, reliquaries, chalices, tabernacles, sacred book covers, ancient Chinese bronzes, burial accessories, the paraphernalia of tribal rituals, the special equipment worn by participants in the sacred ball game of ancient Mexico, processional images, masks and headdresses, and modern trophies and awards.

ART FOR PSYCHOTHERAPY

SECTION FIVE: SKETCHES & DRAWINGS IN ANALYTICAL CONCEPTS

Schematic comparisons

If one schematically compares the three principal tendencies in psychotherapy (Freudian, Jungian, Adlerian) with regard to the direction in which their central thought leads, one could say:-

The analytical method of Sigmund Freud looks for the causae efficientes, the causes of the later behavioural disturbances. Alfred Adler considers and treats the initial situation with regard to a causa finalis and both see in the drives the causae materiales.

In Carl Gustav Jung's case the term 'synthesis' is based on his abandonment of the causal thinking of the alternative psychological methods of treatment. Jungian psychotherapy, therefore, is not an analytical procedure in the usual meaning of this term.

Whatever the differences among Freud's, Jung's and Adler's extensive works on the therapeutic methodologies; scientists, artists, thinkers and practitioners accept the great importance of Freud's and Jung's studies for medicine, psychology, anthropology, religion, art, history, literature, and a plethora of other subjects.

It was in 1960 when I came across the concepts and application of psychotherapy, psycho-analysis, psychiatry, and mental health in general. Having previously spent four years in a general hospital, I had two years in which to finalise my thesis for my Doctorate in Psychology. I had an assortment of cases to assess

ART FOR PSYCHOTHERAPY

and report on successes and the failures of the medical treatment of mental patients as administered at the time.

It all happened in the boundaries of five square miles on a hill in Surrey, U.K. where the psychiatric hospital kept 2,500 patients within its walls and another 2,000 people involved in giving various types of services.

My critique on such methods of treatment made me biased toward the use of medical-less treatment and the best available type of psychotherapy. At that time it was that of the Jungian Psychotherapy and Freud's psycho-analysis.

Fifty years later I visited the site and there was no psychiatric hospital. The construction of suburbia houses was taking place instead. The sufferings of people were demolished together with the buildings and the methods of the somewhat cruel way of treating people.

Interpreting Freudian Concepts

More than seventy-two years have passed since the death of Sigmund Freud (1856-1939), the Austrian psychiatrist and medical consultant who invented the use of morphine and proceeded to establish himself as the father of Psycho-analysis. The agora is still in full session for the debating as to whether he was a genius and if his works had a fruitful impact on the treatment of neurosis and society at large.

Conscious and Subconscious

Sigmund Freud proposed the idea of conscious and subconscious mind. He began psycho-analysis and

ART FOR PSYCHOTHERAPY

proposed theories of infantile sexuality and their effects on adult life. His volume of work on The Interpretation of Dreams influenced the world of art, and many authors based their novels on his writings.

Copernicus of the Mind

At the peak of his career, Sigmund Freud was named as the Copernicus of the Mind. Inspired by Goethe's essay on Nature, he studied medicine in Vienna, but original work in physiology delayed his graduation. He then studied and specialised in neurology, spurred on by physician Breuer, that hysteria can be cured by making a patient recall painful memories under hypnosis, studied under Charcot in Paris and changed over from neurology to psychopathology. To appease his frowning colleagues in Vienna, he published on his return two strictly neurological studies on aphasia and cerebral paralysis, before risking with Breuer, the joint publication of Studien über Hysterie.

Hypnosis

Finding hypnosis inadequate, Freud gradually substituted the method of 'free association'. Allowing the patient to ramble on with his or her thoughts when in a state of relaxed consciousness and, interpreting the data, an abundance of childhood and dream recollections. He became convinced, despite his own puritan sensibilities, of the fact of infantile sexuality. This became the basis of his theory and cost him his friendship with Breuer, lost him many patients and isolated him from the always conservative medical profession.

ART FOR PSYCHOTHERAPY

Repression

Thereafter, he worked alone, publishing many papers and books, which included his work on dreams, which showed that dreams, like neuroses, are disguised manifestations of repressed wishes of sexual origin. Repression, which differs profoundly from mere conscious suppression, Freud explained by reference to a vast reservoir of subconscious, irrational mental activity, the Id, comprising the crude appetites and impulses, loves and hates, particularly those connected with what he termed the Oedipus complex, the infant's craving for exclusive possession of the parent of the opposite sex.

Impulses

These impulses, at variance with civilised behaviour, are repressed by the ego, a portion of the id which at an early stage has become differentiated from it. At a later stage, the super-ego (conscience) develops out of the ego, determining what is acceptable to the ego and what must be repressed.

Repressions

Repressions disappear from consciousness but live on in the id. In sleep or in day-dreaming the ego relaxes its control and the repressed impulses may succeed in pushing themselves into consciousness, but not until the reduced powers of the ego have exercised a censorship, by distorting the unacceptable character of the dream material into something meaningless but acceptable.

ART FOR PSYCHOTHERAPY

Psycho-analysis seeks to uncover these repressions of what they are and replace them by acts of judgement.

Conscious, Ego, Sub-conscious

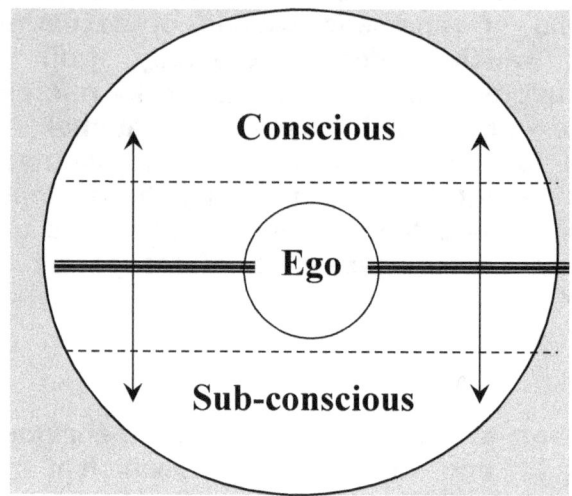

Psychoanalytical Society

In 1902 Sigmund Freud was appointed to an extraordinary professorship, despite previous academic anti-Semitism in Vienna and began to gather disciples, who formed the original 'Psychological Wednesday Society'. Out of this grew the Psychoanalytical Society and in 1910, with Jung as first President, the International Psychoanalytical Association, which included such names as Adler, Jung, Steckel, Rank, Eitingon, Abraham, Ferenczi, Jones, and Brill.

ART FOR PSYCHOTHERAPY

Jung and Adler

Further works, writings and essays by Freud met with heated, incomprehensive opposition and was not before 1930, when Freud was awarded the Goethe prize, that his efforts no longer aroused active opposition from public bodies. This award was bestowed to Freud after Jung and Adler diverged from the Freudian theory by seeking to remove the central emphasis on sexuality.

Breaking up

Adler, who broke with Freud in 1911, developed a psychology of the ego later known as Individual Psychology, and Jung, who followed in 1913, developed a highly complex system of basic human types and the 'collective subconscious, which later on were known as the psychotherapy of Jung. Thereafter, Ernest Jones formed a committee of senior collaborators pledged to uphold the basic Freudian conceptions. Psychoanalysis thenceforth was a creed as well as a science.

Hitler

In 1933, Hitler banned psycho-analysis. After Austria was overrun, Freud and his family were extricated from the hands of the Gestapo by diplomatic representations and allowed to emigrate after Freud had signed a document to the effect that no pressure had been placed upon him by the Nazi government. He settled in Hampstead, London, U.K., and died there in September 23, 1939, from cancer of the jaw which had troubled him for sixteen years.

ART FOR PSYCHOTHERAPY

Royal Society

Freud's work affected a profound revolution in man's attitude towards and comprehension of his mental processes, constituting, after Copernicus and Darwin, 'a third blow to man's self-esteem'. On his eightieth birthday, Thomas Man delivered an address to his honour and he was elected a corresponding member of the Royal Society.

Writings

Other important writings are a work on humour, Totem and Tabu, Beyond the Pleasure Principle, Ego and I, The Future of an Illusion... He collaborated with Albert Einstein in Why War? in 1933.

Inter-relationship of Sciences

For those of us who studied, practised, applied a variety of methods, and inter-related the sciences have no doubt as to the effect on everyday life and the way of living in Western societies. The terminology established by Sigmund Freud is present in the vocabulary of everyday life and the interpretation of his Eros concept is still predominant in the advertising media professions.

ART FOR PSYCHOTHERAPY

Complex

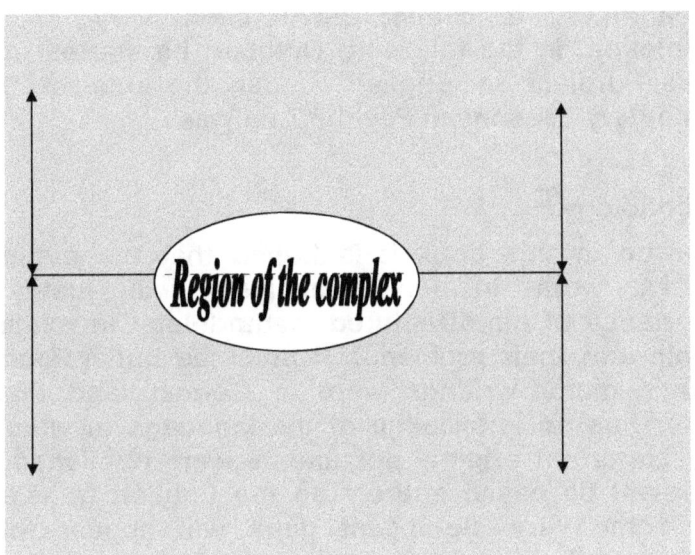

Listening

Methods used in treating people, no matter what therapists maintain, they still insert Freud's principles in listening and transference. Although 'listening' is as old as the human race, to keep quiet when the patient talks and to accept that there is a problem, it is very much a psycho-analytical established way to assist in lifting inhibitions.

ART FOR PSYCHOTHERAPY

London Work

Freud began writing his last work in London on July 1938. A few weeks later he broke it off at a point where, to all appearances, it cannot have been very far from completion. In the following October, he started upon a similar project in English, under the title of 'Some Elementary Lessons in Psycho-Analysis'.

Methodology

Later on in this book it is hoped that the outline on Freud's work in Psycho-analysis will justify the importance of his structured methodology in relating to people and their problems. It must be understood that Freud's major writings were in German and that the author has no knowledge of the language in which the very important original documents were published. This book will be based entirely on the English translations and some very significant gaps will appear for the modern professional psychotherapists.

Doctrines

Nevertheless, it is hoped that the inclusions of Sigmund Freud's series of instructions are clear in this book. Not forgetting that the aim of this book is to bring together the doctrines of psycho-analysis in a précis – in the most concise form and in positive terms. Naturally, its intention is not to compel belief or to establish conviction.

ART FOR PSYCHOTHERAPY

Observations and Experiences

The teachings of psycho-analysis, Freud's and the post-Freudians are based upon an incalculable number of observations and experiences, and no one who has not repeated these observations upon himself or upon others is in position to arrive at an independent judgement of it.

Oedipus

Those with studies in classical literature, mainly the Hellenic period of the Athenian golden age will know that the psycho-analytical terms used are derived and that the Freudian processes are based on the Grecian philosophical and mythological writings. We all speak of the Oedipus complex, Ego, Superego, and conflicts which are extracts from the few surviving classical tomes of the Greek novels.

Philosophical Thought

By now, it is emphatically clear that psycho-analysis makes a basic assumption; the discussion of which falls within the sphere of philosophical thought, but the justification of which lies in its attempt to treat the human psychê.

Psychoanalysis in the Sphere of Philosophy

Psycho-analysis makes a basic assumption, the discussion of which falls within the sphere of philosophical thought, but the justification of which lies in its results. Two things are known concerning what is called psychê and mental life: firstly, its bodily organ and

ART FOR PSYCHOTHERAPY

scene of action, the brain (or nervous system), and secondly, the acts of consciousness, which are immediate data and cannot be more fully explained by any kind of description.

Unknown Points

Everything that lies between these two terminal points is unknown to us and, so far as we are aware, there is no direct relation between them. If it existed, it would at the most afford an exact localisation of the processes of consciousness and would give us no help towards understanding them.

Hypotheses

The two hypotheses start out from these ends or beginnings of our knowledge. The first is concerned with localisation. It is assumed that mental life is the function of an apparatus to which we ascribe the characteristics of being extended in space and being made up of several portions, which we imagine, as being like a telescope or microscope or something of the sort. The consistent carrying through of conception of this kind is a scientific novelty, even though some attempts in that direction have been made previously.

Individual Development

Scientists have arrived at our knowledge of this physical apparatus by studying the individual development of human beings. To the oldest of these mental provinces or agencies the name of Id is given. The id contains

ART FOR PSYCHOTHERAPY

everything that is inherited, that is fixed in the constitution, therefore the instincts, which originate in the somatic organisation and which find their first mental expression in forms unknown to us.

Conscious, Preconscious, Ego, and Subconscious

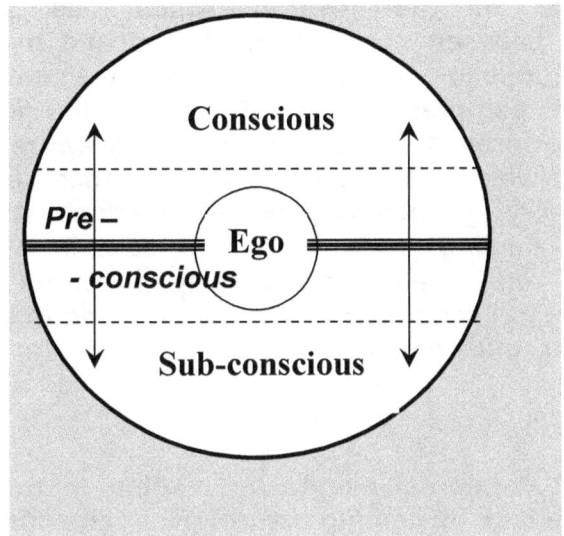

Mental Regions

Freud maintains that under the influence of the external world, one portion of the id has undergone a special development. From what was originally a cortical layer, provided with organs for receiving stimuli and with the apparatus for protection against excessive stimulation, a

ART FOR PSYCHOTHERAPY

special organisation has arisen, which henceforward acts as an intermediary between the id and the external world. This region of the mental life has been given the name of Ego.

Ego

The principal characteristics of the ego are these. In consequence of the relation which has already established between sensory perception and muscular action, the ego is in control of voluntary movement. It has the task of self-preservation. Regarding the external events, it performs that task by becoming aware of the stimuli from without, by sorting up experiences of them (in the memory), by avoiding excessive stimuli (through flight), by dealing with moderate stimuli (through adaptation) and finally, by learning to bring about appropriate modifications in the external world to its own advantage (through activity).

Id

Regarding the internal events, in relation to the id, it performs the task by gaining control over demands of the instincts, by deciding whether they shall be allowed to obtain satisfaction, by postponing that satisfaction to times and circumstances favourable in the external world, or by suppressing their excitation completely. Its activities are governed by consideration of the tensions produced by stimuli present within it, or introduced into it.

ART FOR PSYCHOTHERAPY

Tensions

The raising of these tensions is in general felt as un-pleasure and their lowering as pleasure. It is possible that what is felt as pleasure and un-pleasure is not the absolute degree of the tensions, but something in the rhythm of their changes. The Ego pursues pleasure and seeks to avoid un-pleasure. An increase in un-pleasure that is expected and foreseen is met by signal of anxiety; the occasion of this increase, whether it threatens from without or within, is called danger.

Ego Withdrawal

From time to time the ego gives up its connection with the external world and withdraws into the state of sleep, in which its organisation undergoes far-reaching changes. It may be inferred from the state of sleep that, that organisation consists in a particular distribution of mental energy.

Dependence

The long period of childhood, during which the growing human being lives in dependence upon its parents, leaves behind it a precipitate, which forms within his ego a special agency in which this parental influence is prolonged. It has received the name of Super-ego. In so far as the super-ego is differentiated from the ego or opposed to it, it constitutes a third force which the ego must take into account.

ART FOR PSYCHOTHERAPY

Id, Super-ego, and Reality

Thus, an action by the ego is as it should be if it satisfies simultaneously the demands of the id, of the super-ego and of reality, that is to say if it is able to reconcile their demands with one another. The details of the relation between the ego and the super-ego become completely intelligible if they are carried back to the child's attitude towards his parents. The parents' influence naturally includes not merely the personalities of the parents themselves, but also the racial, national, and family traditions handed on through them as well as the demands of the immediate social milieu which they represent.

Substitutes

In the same way, an individual's super-ego in the course of his development takes over contributions from later successors and substitutes of his parents, such as teachers, admired figures in public life, or high social ideals. It will be seen that, in spite of their fundamental difference, the id and the super-ego have one thing in common; they both represent the influences of the past (the id the influence of heredity, the super-ego essentially the influence of what is taken over from other people), whereas the ego is principally determined by the individual's own experience, that is to say by accidental and current events.

Physical Apparatus

This general pattern of a physical apparatus may be supposed to apply equally to the higher animals which

ART FOR PSYCHOTHERAPY

resemble man mentally. A super-ego must be presumed to be present wherever, as in the case of man, there is long period of dependence in childhood. The assumption of distinction between ego and id cannot be avoided.

Region of Complex

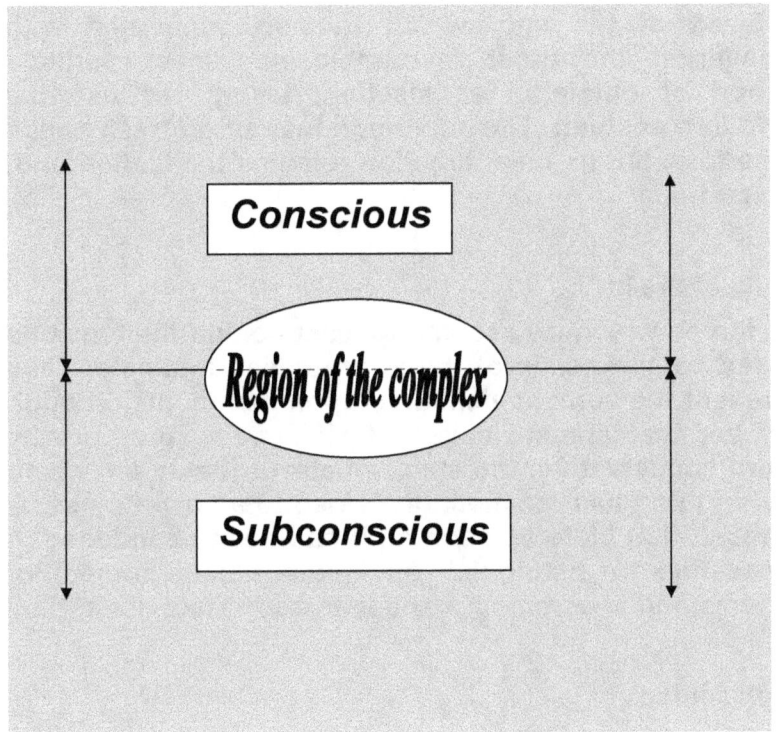

ART FOR PSYCHOTHERAPY

Theory of the Instincts

The power of the id expresses the true purpose of the individual organism's life. This consists in the satisfaction of its innate needs. No such purpose as that of keeping itself alive or of protecting itself from dangers by means of anxiety can be attributed to the id. That is the business of the ego, which is also concerned with discovering the most favourable and least perilous method of obtaining satisfaction, taking the external world into account. The super-ego may bring fresh needs to the fore, but its chief function remains the limitation of satisfactions.

Needs of the Id

The forces which we assume to exist behind the tensions caused by the needs of the id are called instincts. They represent the somatic demands upon mental life. Though they are the ultimate cause of all activity, they are by nature conservative; the state, whatever it may be, which a living thing has reached, gives rise to a tendency to re-establish that state as soon as it has been abandoned. It is possible to distinguish an indeterminate number of instincts and in common practice this is in fact done.

Displacement

For a few of us Analysts, however, the important question arises whether we may not be able to derive all of these various instincts from a few fundamental ones. We found that instincts can change their aim (by displacement) and also that they can replace one another – the energy of

ART FOR PSYCHOTHERAPY

one instinct passing over another. This latter process is still insufficiently understood. After long doubts and vacillations we have decided to assume the existence of only two basic instincts, Eros (Sex) and Thanatos (Death - the destructive instinct).

Self-preservation

The contrast between the instincts of self-preservation and of the preservation of the species, as well as the contrast between ego-love and object-love, falls within the bounds of Eros. The aim of the first of these basic instincts is to establish over greater unities and to preserve them, thus in short, to bind together; the aim on the second, on the contrary is to undo connections and so destroy things. It may be supposed that the final aim of the destructive instinct is to reduce living things to an inorganic state. For this reason it is also called Thanatos, the death instinct.

Eros

If we suppose that living things appeared later than inanimate ones and arose out of them, then the death instinct agrees with the formula that was stated, to the effect that instincts tend towards a return to an earlier state. It will be an impossibility to apply this formula to the love instinct, Eros. That would be to imply that living substance had once been a unity but had subsequently been torn apart and was now tending towards re-union.

ART FOR PSYCHOTHERAPY

Opposing Forces

In biological functions the two basic instincts work against each other or combine with each other. Thus, the act of eating is a destruction of the object with the final aim of incorporating it, and the sexual act is an act of aggression having as its purpose the most intimate union. This interaction of the two basic instincts with and against each other gives rise to the whole variegation of the phenomena of life. The analogy of out two basic instincts extends from the region of animate things to the pair of opposing forces (attraction and repulsion) which rule in the inorganic world.

Sexual Aggressiveness

Modifications in the proportions of the fusion between the instincts have the most noticeable results. A surplus of sexual aggressiveness will change a lover into a sexual murderer, while a sharp diminution in the aggressive factor will lead to shyness or even impotence.

Restrictions

There can be no question of restricting one or the other of the basic instincts to a single region of the mind. They are necessarily present everywhere. We may picture an initial state of things by supposing that the whole available energy of Eros, to which the name of libido is given, is present in the as yet undifferentiated ego-id and serves to neutralise the destructive impulses which are simultaneously present.

ART FOR PSYCHOTHERAPY

Libido

There is no term analogous to 'libido' for describing the energy of the destructive instinct. It becomes relatively easy to follow the later vicissitudes of the libido; but this is more difficult with the destructive instinct.

Psychê

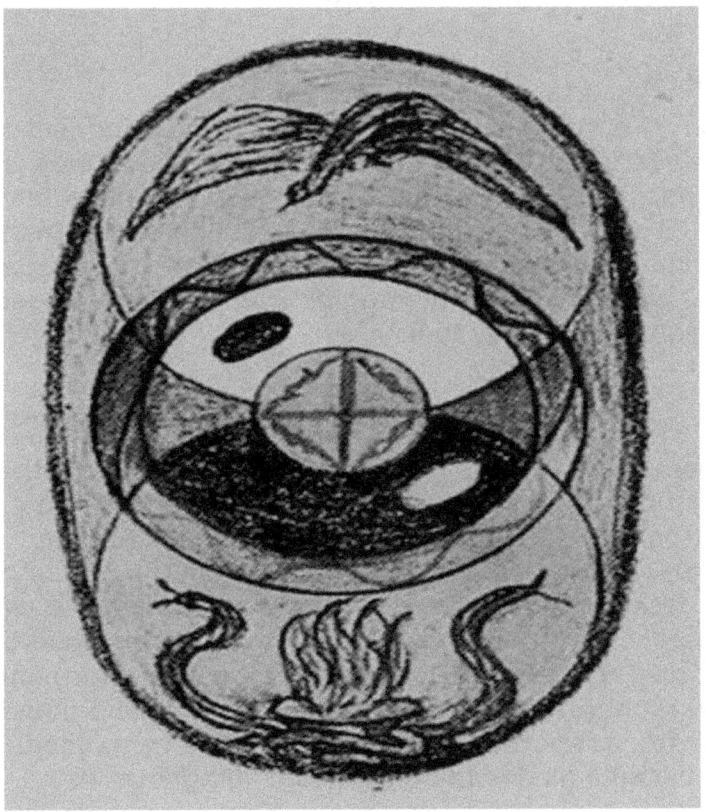

ART FOR PSYCHOTHERAPY

Diversion

So long as the instinct, it remains silent; one comes across it after it has become diverted outwards as an instinct of destruction. That the diversion should occur seems essential for the preservation of the individual; the musculature is employed for the purpose.

Fixation

When the super-ego begins to be formed, considerable amounts of the aggressive instinct become fixated within the ego and operate there in a self-destructive fashion. This is one of the dangers to health to which mankind become subject on their path to cultural development. The holding back of aggressiveness is in general unhealthy and leads to illness.

Self-destructiveness

A person in a fit of rage often demonstrates how the transition from untrained aggressiveness to self-destructiveness is effected, by turning his aggressiveness against himself: he tears his hair or beats his face with his fists – treatment which he would evidently have preferred to apply to someone else. Some portion of self-destructiveness remains permanently within, until it at length succeeds in doing the individual to death, not perhaps until his libido has been used up or has become fixated in some disadvantageous way.

ART FOR PSYCHOTHERAPY

Internal Conflicts

Thus it may in general be suspected that the individual dies of his internal conflicts but that the species dies of its unsuccessful struggle against the external world, when the latter undergoes changes of a kind that cannot be dealt with by the adaptations which the species has acquired.

Narcissistic

It is difficult to say anything of the behaviour of the libido in the id and in the super-ego. Everything that is known about it relates to the ego, in which the whole available amount of libido is at first stored up. This state of things absolute is called primary narcissism. It continues until the ego begins to cathect (from the Greek cathexis) the preservations of objects with libido – to change narcissistic libido into object libido.

Libidinal Cathexes

Throughout life the ego remains the great reservoir from which libidinal cathexes are sent out on to objects and into which they are also once more withdrawn, like the pseudopodia of a body of protoplasm. It is only when someone is completely in love that the main quantity of libido is transferred on to the object and the object to some extent takes the place of the ego.

A characteristic of libido which is important in life in its mobility, the case with which it passes from one object to another. This must be contrasted with the fixation of libido to particular objects, which often persists through life.

ART FOR PSYCHOTHERAPY

Erotogenic Zones

There can be no question that the libido has somatic sources that it streams into the ego from various organs and parts of the body. This is most clearly seen in the case of the portion of the libido which, from its instinctual aim, is known as the sexual excitation. The most prominent of the parts of the body from which this libido arises are described by the name of erotogenic zones, though strictly speaking the whole body is an erotogenic zone.

Eros

The greatest part of what we know about Eros, which is about the exponent, the libido has been gained from the study of the sexual function, which indeed, in the popular view co-incites with Eros.

Sexual Impulse

A picture has been formed of the way in which the sexual impulse, which is destined to exercise a decisive influence on our life, gradually develops out of successive contributions from a number of component instincts, which represent particular erotogenic zones.

ART FOR PSYCHOTHERAPY

Super-ego, Ego, Instincts

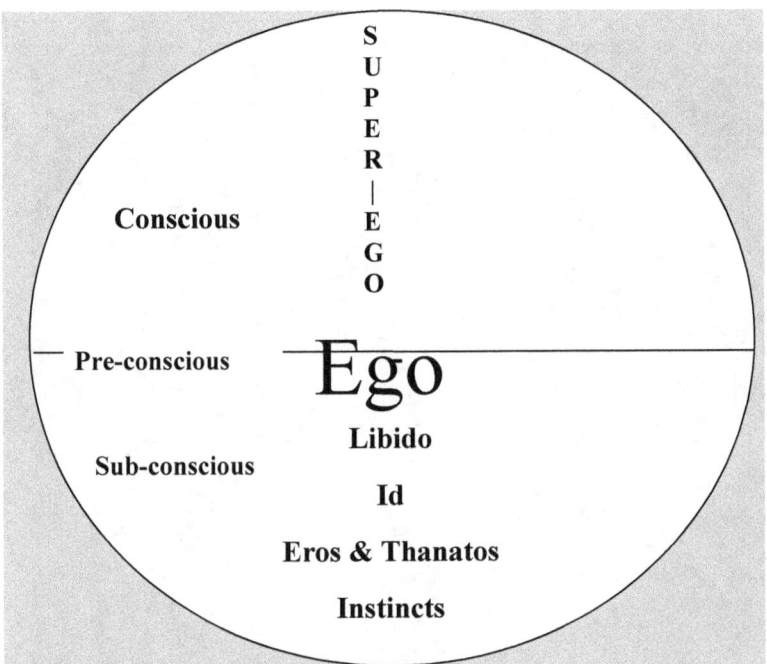

Development of the Sexual Function

According to the popular view, human sexual life consists essentially in the impulse to bring one's own genitals into contact with those of someone of the opposite sex. With this are associated, as accessory phenomena and introductory acts, kissing this extraneous body, looking at and touching it. This impulse is supposed to make its appearance in puberty, that is, at the age of sexual maturity, and to serve the purposes of reproduction.

ART FOR PSYCHOTHERAPY

Fruitful Relationship

ART FOR PSYCHOTHERAPY

Attractions

Nevertheless, certain facts have always been known that fail to fit the narrow framework of this view. It is a remarkable fact that there are people who are only attracted by the persons and genitals of members of their sex. It is equally remarkable that there are people whose desires behave in every way like sexual ones, but at the same time entirely disregard the sexual organs or their normal use; people of this kind are known as 'perverts'. Finally, it is striking that many children (who are on that account regarded as degenerates) take a very early interest in their genitals and show signs of excitation in them.

Sexuality

It may well be believed that psycho-analysis provoked astonishment and denials when, partly upon the basis of these three neglected facts, it contradicted all the popular opinions upon sexuality.

Its principal findings are as follows:

- Sexual life does not begin only at puberty, but starts with clear manifestations soon after birth.
- It is necessary to distinguish sharply between the concepts of 'sexual' and 'genitals'. The former is the wider concept and includes many activities that have nothing to do with the genitals.
- Sexual life comprises the function of obtaining pleasure from zones of the body, a function which is subsequently brought into the service of that of reproduction. The two functions often fail to coincide completely.

ART FOR PSYCHOTHERAPY

Bodily Activity

The chief interest is naturally focused upon the first of these assertions, the most unexpected of all. It has been found that in early childhood there are signs of bodily activity of which only ancient prejudice could deny the name of sexual, and which are connected with mental phenomena that we come across later in adult love, such as fixation to a particular object, jealousy, and so on.

Phenomena

It is further found that these phenomena which emerge in early childhood form part of a regular process of development, that they undergo a steady increase and reach a climax towards the end of the fifth year, after which there follows a lull, progress is at a standstill and much is unlearnt and undone. After the end of latency, as it is called, sexual life is resumed with puberty, or as we might say, it has a second efflorescence.

Infantile Amnesia

Here we come upon the fact that the onset of sexual life is dysphasic, that it occurs in two waves; this is unknown except in man and evidently has an important bearing upon his genesis. It is not a matter of indifference that, with few exceptions, the events of the early period of sexuality fall a victim to infantile amnesia.

ART FOR PSYCHOTHERAPY

Aetiology

Our understanding of the aetiology of the neuroses and the technique of analytic therapy are derived from these views; and the tracing of the process of development in this early period has also provided evidence for yet other conclusions.

Libidinal Demands

The first organ to make its appearance as an erotogenic zone and to make libidinal demands upon the mind is, from the time of birth onwards, the mouth. To begin with, all mental activity is centred upon the task of providing satisfaction for the needs of that zone. In the first instance, of course, the latter serves the purposes of self-preservation by means of nourishment; but physiology should not be confused with psychology.

Obstinate Persistence

The baby's obstinate persistence in sucking gives evidence at an early stage of a need for satisfaction which, although it originates from and is stimulated by the taking of nourishment, nevertheless seeks to obtain pleasure independently of nourishment and for that reason may and should be described as 'sexual'.

Sadistic Impulses

Sadistic impulses already begin to occur sporadically during the oral phase along with the appearance of the teeth. Their extend increases greatly during the second phase, which we describe as the sadistic-anal phase, because satisfaction is then sought in aggression and in excretory function. We justify our inclusion of aggressive impulses in the libido by supposing that sadism is an instinctual fusion of purely libidinal and purely destructive impulses, a fusion which thenceforward persists without interruption.

Phallic Phase

The third phase is the so-called phallic one, a forerunner of the final shape of sexual life, and already greatly resembles it. It is to be noted that what comes in question at this stage is not the genitals of both sexes but only those of the male, the phallus. The female genitals long remain unknown: in the child's attempt at understanding sexual processes, he pays homage to the venerable cloacal theory, a theory which has a genetic justification.

Penis Presence

With the phallic phase and in the course of it, the sexuality of early childhood reaches its height and approaches its decline. Thenceforward boys and girls have different histories. To begin with, both place their intellectual activity at the service of sexual research; both start off from the presumption of the universal presence of the penis.

Oedipus Phase

But now the paths of the sexes divide. The boy enters the Oedipus phase; he begins to manipulate his penis, and simultaneously has fantasies of carrying out some sort of activity with it in relation to his mother; but at last, owing to the combined effect of a threat of castration and the spectacle of women's lack of penis, he experiences the greatest trauma of his life, and this introduces the period of latency with all its attendant consequences.

Electra Phase

The girl, after vainly attempting to do the same as the boy, comes to recognise her lack of a penis or rather the inferiority of her clitoris, with permanent effects upon the development of her character; and as a result of this first disappointment in rivalry, she often turns away altogether from sexual life.

Simultaneous Presence

It would be a mistake to suppose that these three phases succeed on another in a clear-cut fashion: one of them may appear in addition to another, they may be present simultaneously.

Independent Pleasure

In the earlier phases the separate component instincts set about their pursuit of pleasure independently of one another; in the phallic phase there are the first signs of an organisation which subordinates the other trends to the primacy of the genitals and signifies the beginning of a co-ordination of the general pursuit of pleasure into the sexual function.

ART FOR PSYCHOTHERAPY

Puberty

The complete organisation is not attained until puberty, in the fourth or genital phase.

A state of affairs is then established in which:
- Many earlier libidinal cathexes are retained,
- Others are included in the sexual function as preparatory or auxiliary acts, their satisfaction producing what is known as fore-pleasure,
- Other tendencies are excluded from the organisation, and are either entirely suppressed (repressed) or are employed in the ego in some other way, forming character-traits or undergoing sublimation with a displacement of their aims.

Inhibitions

This process is not always carried out perfectly. Inhibitions in the course of its development manifest themselves as to various disturbances of sexual life. Fixations of the libido to conditions at earlier phases are then found, the trend of which, moving independently of the normal sexual aim, is described as perversion.

Homosexuality

One example of an inhibition in development of this kind is homosexuality, if it is manifest. Analysis shows that in every case a homosexual attachment to an object has at one time been present and in most cases has persisted in a latent condition.

ART FOR PSYCHOTHERAPY

Cathexes

The condition is complicated by the fact that the processes necessary for bringing about a normal outcome are not for the most part either completely present or completely absent; they are as a rule partially present, so that the final result remains dependent upon quantitative relations. Thus genital organisation will be attained, but it will be weakened in respect of those portions of the libido which have not proceeded so far but have remained fixated to pre-genital objects and aims.

Tendency

Such weakening shows itself in a tendency, if there is an absence or genital satisfaction, or if there are difficulties in the real world, for the libido to return to its earlier pre-genital cathexes (i.e. to regress).

Phenomenology

During the study of the sexual functions it has been possible to gain a first, preliminary conviction, or rather suspicion, of two pieces of knowledge which will later be found to be important over the whole of the field of analysis.

Dynamics and Economics

Firstly, the normal and abnormal phenomena that we observe (that is, the phenomenology of the subject) require to be described from the point of view of dynamics and of economics (i.e., in this connection, from

ART FOR PSYCHOTHERAPY

the point of view of the quantitative distribution of the libido).

Aetiology of the Disturbances

Secondly, the aetiology of the disturbances which we are studying is to be found in the developmental history of the individual, that is to say, in the early part of his life.

Antithesis

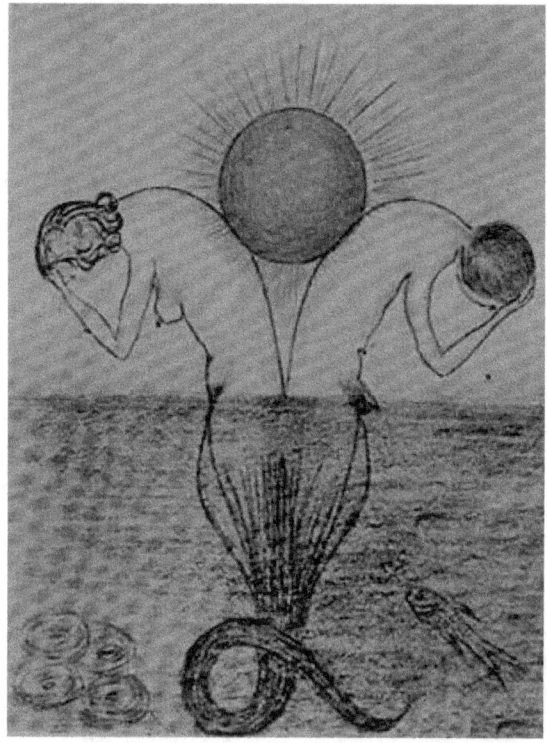

ART FOR PSYCHOTHERAPY

Mental Structure

The structure of the psychical apparatus and the energies or forces which are active in it, also the way in which these energies, and principally the libido, organise themselves into a physiological function (which serves the purpose of the preservation of the species), were described in the previous chapter.

Mental Life

There was nothing in all this to exemplify the quite peculiar character of what is mental, apart of course, from the empirical fact that this apparatus and these energies underlie the functions which are called mental life. Mental life, therefore, is a unique characteristic, which according to a widely held opinion, actually co-insides with the exclusion of all else.

Consciousness

The starting point for this investigation is provided by a fact without parallel, which defies all explanation or description – the fact of consciousness. Nevertheless, if anyone speaks of consciousness, we know immediately and from out of our own most personal experience what is meant by it.

Consciousness Assumption

Many people, both inside and outside the science of psychology, are satisfied with the assumption that consciousness alone is mental, and nothing then remains for psychology but to discriminate in the phenomenology of the mind between perceptions, feelings, intellective processes, and volitions.

Conscious Processes

It is generally agreed, however, that these conscious processes do not form unbroken series which are complete in themselves; so that there is no alternative to assuming that there are physical or somatic processes which accompany the mental ones and which must admittedly be more complete than the mental series, since some of them have conscious processes parallel to them but others have not.

Somatic Processes

It thus seems natural to lay the stress in psychology upon these somatic processes, to see in them the true essence of what is mental and to try to arrive at some other assessment of the conscious processes. The majority of philosophers, however, as well as many other people, dispute this position and declare that the notion of a mental thing being subconscious is self-contradictory.

Fundamental Hypothesis

But it is precisely this that psycho-analysis is obliged to assert, and this is its second fundamental hypothesis. It explains the supposed somatic accessory processes as being what is essentially mental and disregards for the moment the quality of consciousness. It does not stand alone in this opinion. Many thinkers, such as Theodor Lipps, have made the same assertion in the same words.

ART FOR PSYCHOTHERAPY

Psycho-analysis and Philosophy Disputes

Now it might appear as though this dispute between psycho-analysis and philosophy was only concerned with a trifling matter of definition; the question whether the name 'mental' should be applied to one or another series of phenomena. Actually, this step has been of great importance.

Natural Science

Whereas psychology of consciousness never went beyond this broken sequence of events which was obviously dependent upon something else, the other view that what is mental is in itself subconscious, enabled psychology to take its place as a natural science.

Apparatus

Every science is based upon observations and experiences arrived at through the medium of our physical apparatus. But since our science has as its subject that apparatus itself, the analogy ends here. Observations are made through the medium of the same perceptual apparatus, precisely by the help of the breaks in the series of conscious mental events, since we fill in the omissions by plausible inferences and translate them into conscious material.

Conscious Events

In this way we construct a series of conscious events complementary to the subconscious mental processes.

ART FOR PSYCHOTHERAPY

The relative certainty of our mental science rests upon the binding force of these inferences. Anyone who goes deeply into the subject will find that our technique holds its ground against any criticism.

Mental Processes

Thus, three qualities have been attributed to the mental processes: they are conscious, preconscious, or subconscious. The division between the three classes of material which have these qualities are neither absolute nor permanent. What is preconscious becomes conscious and what is subconscious can be made conscious, although in the process we may have the impression that we are overcoming very strong resistances.

Internal Resistances

From this we may infer that the maintenance of certain internal resistances is a sine quâ non of normality. A lowering of resistance of this sort, with a consequent pressing forward of subconscious material takes place regularly in the state of sleep and thus brings about a necessary precondition for the formation of dreams.

Perceptions

The process of a thing becoming conscious is above all linked with the perceptions which our sense organs receive from the external world. From the topographical point of view, therefore, it is a phenomenon which occurs in the outermost cortex of the ego.

ART FOR PSYCHOTHERAPY

Peremptory Influence

It is true that we also receive conscious information from outside of the body; the feelings, which actually exercise a more peremptory influence upon our mental life than external perceptions. In certain circumstances the sense organs themselves transmit feelings, sensations, or pain, in addition to the perceptions which are specific to them.

Energy

It is assumed, as the other natural sciences teach, that in mental life some kind of energy is at work; but there are no data which enable us to come nearer to knowledge of it by an analogy with other forms of energy.

Cathexes and Hyper-cathexes

We seem to recognise that nervous or physical energy exists in two forms, one freely mobile and the other, by contrast; we speak of cathexes and hyper-cathexes of the material of the mind and even venture to suppose that a hyper-cathexis brings about a sort of synthesis of different processes.

Primary Process

Behind all these uncertainties there lies one new fact, the discovery of which we owe to psycho-analytic research. That processes in the subconscious or in the id obey different laws from those in the preconscious ego. These laws in their totality are called the primary process, in

ART FOR PSYCHOTHERAPY

contrast to the secondary process which regulates events in the preconscious or ego.

Dream Interpretation

An investigation of normal, stable states, in which the frontiers of the ego are safeguarded against the id by resistances (or anti-cathexes) and have held firm, and in which the super-ego is not distinguished from the ego because they work together harmoniously – an investigation of this kind will teach us little.

Conflict and Rebellion

The only thing that can help us are state of conflict and rebellion, in which the material in the subconscious id has a prospect of forcing its way into the ego and into consciousness and in which the ego arms itself afresh against the invasion. Only under such conditions can we make observations which will confirm or correct our views upon the two partners.

Dreams Normality

But one nightly sleep is precisely a state of this sort, and consequently our activity during sleep, which we perceive as dreams, is the most favourable object of Freud's study. In this way, too, we can avoid the familiar charge of basing our constructions of the normal life of the mind upon pathological findings; for dreams are regular events in the life of normal people, however much of their characteristics may differ from the productions of the waking existence.

ART FOR PSYCHOTHERAPY

Dreams

Dreams as everyone knows, can be confused, unintelligible or positively senseless, their contents may contradict all that is known of reality, and people behave in them like insane individuals, since, so long as we are dreaming, we attribute objective reality to the material of dreams.

Dream-process

We can find our way towards understanding (or 'interpreting') dreams, if we assume that what we collect as the dream, after we have woken up, is not the true dream-process but only a façade behind which that process lies concealed.

Dream Manifestation

Here we have our distinction between manifest dream-material and latent dream-thoughts. The process which produces the former out of the latter is described as dream-work.

Ego Forcing

The study of dream-work affords an excellent example of the way in which subconscious material from the id forces itself upon the ego, becomes pre-conscious and, owing to the efforts of the ego, undergoes the modification which is called dream-distortion. There are no features of the dream which cannot be explained in this fashion.

ART FOR PSYCHOTHERAPY

Mechanism of Dreams

The mechanism of dream-formation is the same in all conditions. The ego gives evidence of its origin from the id by occasionally ceasing its functions and permitting a reversion to an earlier state of things. It duly brings this about by breaking off its relations with the external world and withdrawing its cathexes from the sense organs.

Superfluity

Since the waking ego controls the power of movement, that function is paralysed in sleep, and accordingly a great part of the inhibitions imposed upon the subconscious id becomes superfluous.

Evidence of Dream-work

The evidence of the share taken by the subconscious id in the formation of reams is abundant and convincing:

- Memory is far more comprehensive in dreams than in waking life. Dreams bring up recollections which the dreamer has forgotten, which are inaccessible to him when he is awake.
- Dreams make an unlimited use of linguistic symbols, the meaning of which is for the most part unknown to the dreamer.
- Memory very often reproduces in dreams impressions from the dreamer's early childhood of which it can definitely be asserted not only that they had been forgotten, but that they had become subconscious owing to repression.

ART FOR PSYCHOTHERAPY

- Dreams bring to light material which could not originate either from the dreamer's adult life or from his forgotten childhood.

Subconscious Work-over

But what makes dreams so invaluable for giving knowledge is the circumstance that, when the subconscious material forces its way into the ego, it carries along with it its own methods of working. Thus dream-work is in its essence a case of a subconscious working-over of pre-conscious thought processes.

Kingdom of the Illogical

The study of dream-work has taught us many other equally remarkable and important characteristics of the processes in the subconscious; but a few of them can still be mentioned. The governing laws of logic have no sway in the subconscious; it might be called the Kingdom of the Illogical.

Compromise

Impulses with contrary aims exist side by side in the subconscious without any call being made for an adjustment between them. Either they have no effect, or no decision is made, but a compromise comes about which is senseless since it embraces mutually exclusive elements.

ART FOR PSYCHOTHERAPY

Wish Fulfilment

In general every dream can be a fulfilment to a wish. Naturally every case is not so simple. Especially in those dreams that arise from residues of the previous day which have not been dealt with and which have merely obtained reinforcement during sleep from the subconscious, it is often hard to detect the subconscious motive force and its wish fulfilment, but it may be assumed that it is always there. The assertion that dreams are wish-fulfilments will easily arouse scepticism when it is remembered how many dreams have a positively painful content and anxiety.

Product of Conflict

It must not be forgotten that dreams are invariably the product of a conflict, that they are a kind of compromise-structure. Something that is a satisfaction for the subconscious id may for that very reason be a cause of anxiety for the ego. Taking into consideration all observations it can be said that every dream is an attempt to put aside a disturbance of sleep by means of a wish-fulfilment. The dream is thus the guardian of sleep.

Subconscious Mechanism

Experience has shown that the subconscious mechanism which is apparent in the study of dram-work and which gives an explanation of the formation of dreams also assists in understanding the puzzling symptoms which attract interest in neurotic and psychotic cases. A dream then is a psychosis, with all the absurdities of delusions and illusions. The necessary condition for the

ART FOR PSYCHOTHERAPY

pathological states can only be a relative or an absolute weakening of the ego, which prevents it from performing its task.

Super-ego Claims

The claims made by the super-ego may become so powerful and so remorseless that the ego may be crippled for its other tasks. It may be suspected that, in the economic conflicts which arise, the id and the super-ego often make common cause against the hard-pressed ego, which, in order to retain its normal state, clings on to reality.

Disturbed Reality

But if the id and the super-ego are too strong, they may succeed in loosening the organisation of the ego and altering it so that its proper relation to reality is disturbed or even abolished.

Jungian Syntheses

The Swiss psychiatrist, Carl Gustav Jung popularised the terms 'introvert and extravert', interpreted the deeper conscious levels in terms of mythology, and established psychotherapy as the treatment of disorders with extensive research in various psychological methods.

ART FOR PSYCHOTHERAPY

Studies and Experience

Professor Jung (1875-1961) studied medicine at Basel, and worked under Bleuler at the Bugholsli clinic at Zurich (1900-1907). He established the term 'complex' in his early studies in word association, and his 1907 publication of The Psychology of Dementia Praecox led to his meeting Sigmund Freud in Vienna. He became Freud's leading collaborator and was elected President of the International Psychoanalytical Association from 1910 to 1914.

Critical of Freud

His independent researches made him increasingly critical of Freud's sexual definition of the libido. His publication of The Psychology of the Unconscious caused a break in 1913. From then onwards he developed his own theories, foremost among which were his description of psychological types ('extraversion/introversion' 1921).

Self –regulating System

His theory of psychic energy emphasised a final point of view as against a purely causal one. His discovery and exploration of the 'collective unconscious', with its 'archetypes' was an impersonal substratum underlying the 'personal unconscious'; the concept of the psyche as a 'self-regulating system' expressing itself in the process of 'individualisation'.

ART FOR PSYCHOTHERAPY

Symbolism

To this latter process Jung devoted most of his latter work, constantly enlarging the scope of his researches; to include the interpretation of the dreams and drawings of patients, the symbolism of religions, myths, historical antecedents as (e.g. alchemy), and even modern physics ('synchronicity').

Influences on Sciences

Thus, Jung's work has become of great importance for medicine, psychology, anthropology, religion, art, history, literature, etc.

Condensed Picture

This presentation of Jung's psychotherapy is intended to give a condensed picture and an introduction to his extensive publications and method of therapy. Above all, to wet the reader's appetite for further interest in Jung's own extraordinarily voluminous works. It is the author's opinion that it would be inappropriate to attempt a description of Jung's forty plus years of intensive research, in a few pages that this book can afford.

Short Task

In short, a practically impossible task. It must necessarily remain a sketch, which the author attempts to organise as simply and clearly as possible, but that must renounce going into profundities or details.

ART FOR PSYCHOTHERAPY

Science not Philosophy

To consider Carl Gustav Jung as another Thinker/Philosopher is still an honour for a man who devoted his time in bringing forward his philosophical thoughts. Lest not ignore the fact that the Jungian psychotherapists, and Jung himself consider their method to be science; neither a school of philosophy, nor a religion.

Term Established

Carl Gustav Jung established the term of Psychotherapy as a part of the wider aspect of his psychological and psychiatric studies. In this manner, the Jung Psychotherapeutic works are divisible into a theoretical part, whose principal headings can be described quite generally as:

- Nature and Structure of the Psyche,
- Laws of the Psychic Processes and Forces,
- The practical part based on these theories, their application, as therapeutic method in the narrower sense.

Philosophical Derivation

If one would reach a correct comprehension of Jung's system, one must first of all accept Jung's standpoint and recognise with him the full reality of the psychic functions. This point of view was, remarkable as it may sound, relatively new at his time. For up to a few decades earlier, the psyche was not considered as independent and subject to its own laws, but was studied and

ART FOR PSYCHOTHERAPY

interpreted through derivation from philosophy, religion or from natural science, so that its true nature could not rightly be discerned.

Psychic Equals Physical

To C G Jung the psychic is no less real than the physical. Though it is not immediately touchable and visible, it is still fully and unambiguously experienceable. Even in the twenty-first century, it is a world in itself – subject to law, structured, and possessed of its special means of expression. All that we know of the world comes to us, as does all knowledge of our own being, through the medium of the psychic, which is therefore one of the most important aspects and conditions of experience.

Interchangeable Subjects

To study it as such was Jung's aim; not however to elevate it as would a mere psychologism to be the sole ground of all knowledge. The psychological, physical, and physico-mathematical standpoints (as well as many others) are interchangeable and can be studied at will according to the problem and the special interests of the enquirer.

Psychological Aspect

Jung took the psychological aspect, leaving the others to persons competent in their fields, drawing however upon his wide acquaintance with psychic reality, so that this theoretical structure is no abstract system created by the speculative intellect but an erection upon the solid ground of experience and resting only on that.

Principles

Its two main pillars were:
- The principle of psychic totality,
- The principle of psychic dynamics.

These two points were elaborated, together with the practical application of the system, in Jung's researches and book publications.

Psyche, Soul, or Mind

By using the term 'psyche' Jung understood not merely what we usually mean by the word 'soul' or 'mind', but the totality of all psychological processes, both conscious and unconscious. That is something broader than and including the soul, which for him constituted only certain limited complex of functions. According to his definitions, the psyche consists of two spheres supplementing one another but opposed in their properties – of consciousness and the so called unconscious.

Ego

The ego has a share of both. The following diagram shows the ego standing between two spheres, which not only supplement but also complement or compensate each other. That is, the dividing line that marks them off from each other in our ego can be displaced in both

ART FOR PSYCHOTHERAPY

directions, as is suggested by the arrows and the dotted lines in the figure.

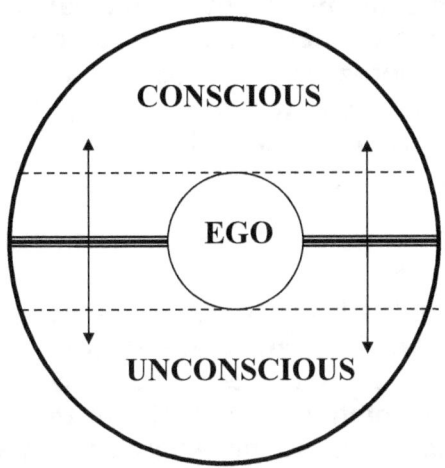

Centre of Reference

The ego itself is not exclusively conscious, but is conceived as a centre of reference for conscious and unconscious psychic contents alike. It forms, as a concept embracing the unitary totality of our psychosomatic beings. It is naturally only expedient of thought and an abstraction that the ego stands exactly in the middle.

Consciousness

Jung defines consciousness as "the function or activity which maintains the relation of the psychic contents to the ego". The next diagram (paragraph 3.7) shows how the sphere of consciousness is surrounded by contents

ART FOR PSYCHOTHERAPY

lying in the unconscious. Here are those contents which have been put aside (for our consciousness can take only a very few contents at once) but which can be raised again at any time into consciousness; furthermore, those which can be repressed because they can be disagreeable for various reasons - i.e., "forgotten, repressed, subliminally perceived, thought, and felt matter of any kind."

Personal Unconscious

This region Jung called it the 'Personal Unconscious' in order to distinguish it from that of the 'Collective Unconscious', as is indicated in the following diagrammatic representation (see 11.7). For the collective part of the unconscious no longer includes contents that are specific for the individual ego and result from the personal acquisitions, but such as result "from the inherited possibility of psychical functioning in general, namely from the inherited brain structure." This inheritance is common to all humanity, perhaps even to the entire animal world, and forms the basis of every individual psyche.

Primal Datum

Further on, Jung maintained that the unconscious is older than consciousness. He added that it is the primal datum out of which 'ever afresh arises'. Thus, consciousness is merely built upon the fundamental psychic activity, which consists in the functioning of the unconscious.

ART FOR PSYCHOTHERAPY

Unconscious Sphere

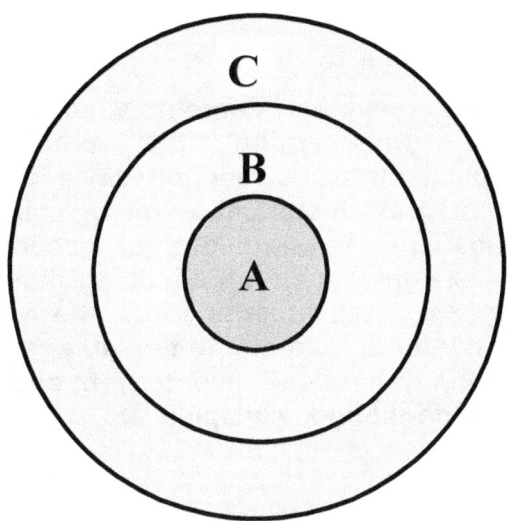

A: The part of the collective unconscious that can never be raised into consciousness,

B: The sphere of the collective unconscious,

C: The sphere of the personal unconscious.

Consciousness Dependent on Unconscious

The notion that man's psychic life is in the main conscious is false, for we spend the greater part of our life in the unconscious: we sleep or daydream... It is incontestable that every important situation in life our consciousness is dependent upon the unconscious. Jung

ART FOR PSYCHOTHERAPY

added that children begin life in an unconscious state and grow into a conscious one.

Unconscious Contents

The unconscious consists of contents which are entirely undifferentiated, representing the precipitate of humanity's typical forms of reaction since the earliest beginnings, apart from historical, ethnological, racial, or other differentiations, in situations of general human character. For example, such situations as those of fear, danger, struggle against superior force, the relations of the sexes, of children to parents, to the father and mother imago, of reaction to hate and love, to birth and death, to the power of the bright and principles, etc.

ART FOR PSYCHOTHERAPY

Psychic Functions

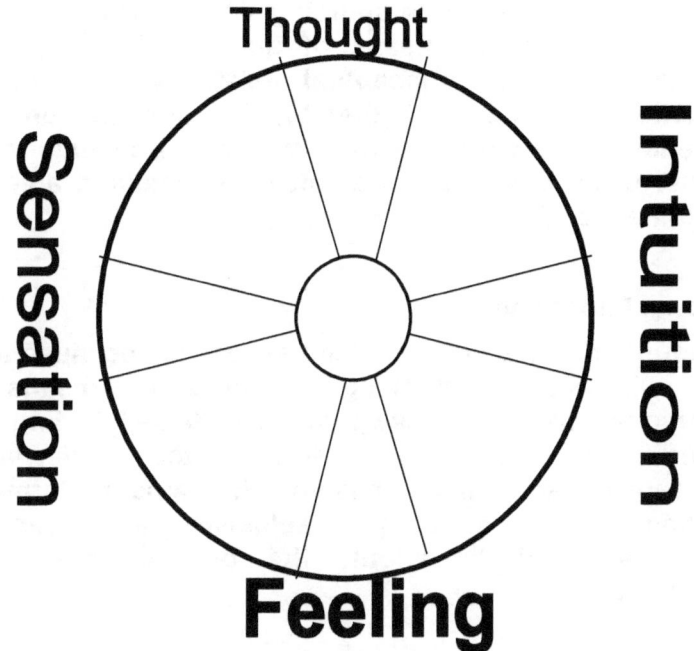

By a psychological function Jung understood a "certain form of activity that remains theoretically the same under varying circumstances and is completely independent of its momentary contents".

ART FOR PSYCHOTHERAPY

Decisive Facts

The decisive fact is not what one thinks, but that one employs one's intellectual function and not one's intuition in receiving and working up contents presented from without or within. Thinking is that function which seeks to reach an understanding of the world and an adjustment to it by means of an act of thought, or cognition, i.e., of conceptual relations and logical deductions. In contrast thereto, the feeling function apprehends the world on the basis of an evaluation by means of the concepts, pleasant or unpleasant, adience or avoidance.

Rational Functions

Both functions are characterised as rational because they work with values: thinking evaluates by means of cognitions from the viewpoint 'true/false, feeling by means of emotions from the viewpoint 'agreeable/disagreeable'. These two fundamental forms of reactions are mutually exclusive as practical determinants of behaviour; the one or the other predominates.

Irrational Functions

The other two functions, sensation and intuition, Jung called the irrational functions, since they circumvent the ratio and work not with judgements but with mere perceptions, without evaluation or interpretation. Sensation perceives things as they are and not otherwise. It is the sense of reality par excellence, what the French call the 'fonction du rèel'.

ART FOR PSYCHOTHERAPY

Inner Perception

Intuition perceives likewise, but less through the conscious apparatus of the senses than through its capacity for an unconscious 'inner perception' of the potentialities in things.

Historical Event

The sensation type will take notice of an historical event in all its details but disregard the psychological context in which it is set; the intuitive, on the contrary, will pass over the details carelessly but perceive without difficulty the inner meaning of the occurrence, its possible relations and consequences.

Dominant Function

Although man possesses constitutionally all four functions, experience shows that it is always only one of these functions with which he orientates himself and adjusts himself to reality. This function becomes the dominant function for adjustment; it gives the conscious attitude its direction and quality and stands constantly at the disposal of the individual conscious will.

Typology

If we wish to give a complete schematic representation of the personality according to Jung's typological system, we can think of introversion-extraversion as constituting a third axis perpendicular to the cross axes of the four functional types. Referring each of the four functions to

ART FOR PSYCHOTHERAPY

both the attitudinal types, we get an eightfold spatial figure. The idea of the quaternity is in fact seldom expressed by the double four, the eight, as well as by the four itself.

Mixed Types

The four functional types, based on the predominance of the one or the other function in the individual are valid in this form only theoretically. In real life they almost never occur pure but more or less as mixed types as is suggested in the following diagram.

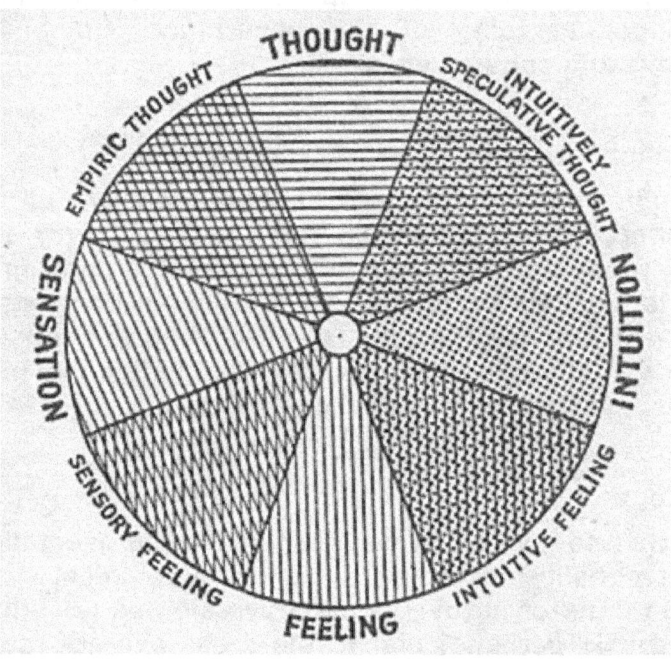

ART FOR PSYCHOTHERAPY

Over-differentiation

The complementary or compensatory relation of the functions to each other is a law inherent in the structure of the psyche. This almost inevitable over-differentiation of the superior function in the course of the years leads nearly always to tensions, which belong to the real problems of the second half of life and whose solution forms one of the principal tasks of this period.

Persona

This specific form of the general psychic behaviour of man with respect to the external world, which Jung called the Persona, is also connected with this over-differentiation.

Sphere of Consciousness

This diagram shows how the whole system of relations through which the psyche manifests itself in relation to the environment shuts off of the ego from the objective world.

ART FOR PSYCHOTHERAPY

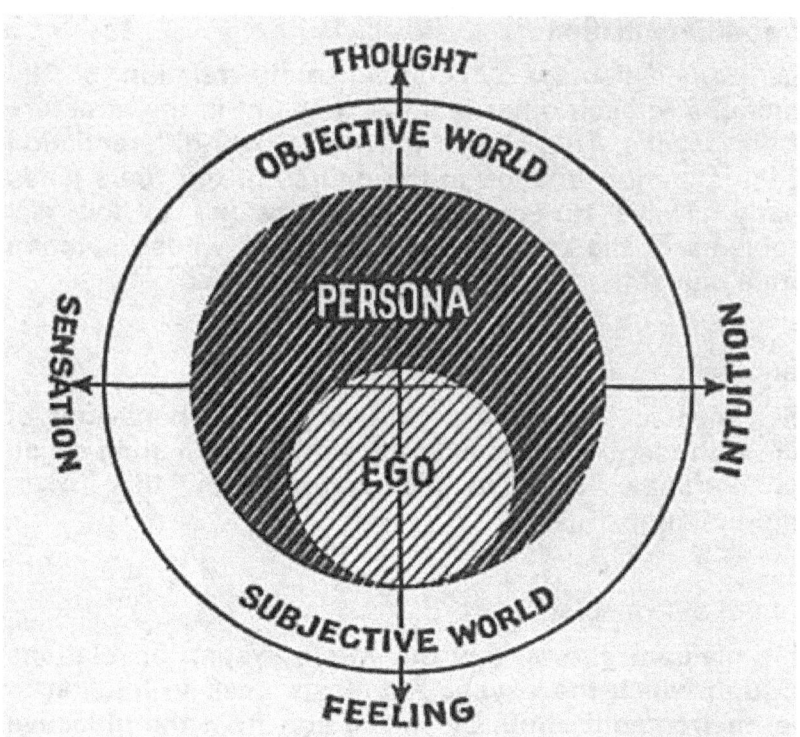

Function-complex

Jung defines the Persona as follows: "The persona is a function-complex which has come into existence for reasons of adaptation or necessary convenience, but by no means is it identical with the individuality. The function-complex of the persona is exclusively concerned with the relation to the object, to the exterior world. The persona is a compromise between the individual and society based on that which one appears to be."

ART FOR PSYCHOTHERAPY

Stigmatisation

Adjustment to the environment can occasionally, however, be attempted not by means of the superior function, as is the rule, but by the inferior. It then succeeds correspondingly unsatisfactory. The persona inevitably appears in this case stigmatised with all the inadequacies that characterise the inferior, undifferentiated function. Such persons not only make an unnatural, artificial impression, but they can easily mislead the psychologically naïve to an entirely false estimate of their real nature.

Personality Inflation

But, not only the bearers and representatives of collective consciousness, the 'big names' attested by community and society, the badges of title, dignities, roles, etc., constitute an attraction and therefore cause an inflation of the personality. Beyond our ego there is not just the collective consciousness of society but also the collective unconscious, our own deep, which conceals equally attractive and imposing figures.

Psychic Health

A well fitting and functional persona is an essential condition for the psychic health and is of the greatest importance if the demands of the environment are to be met successfully.

ART FOR PSYCHOTHERAPY

Character Index

The functional type to which one belongs would be in itself an index to a man's psychological character. It alone, however, will not suffice. In addition his general psychological attitude, i.e., his way of reacting to what meets him from without or within, must be determined.

Extraversion and Introversion

Jung distinguishes two such attitudes: Extraversion and Introversion. They represent orientations that essentially condition all psychic processes, the reaction habitus, namely through which one's way of behaving, of subjectively experiencing, and even of compensating through the unconscious is given.

Habitus, the Central Switchboard

This habitus Jung calls "the central switchboard, from which on the one hand external behaviour is regulated and on the other specific experience is formed." Extraversion is characterised by a positive relation to the object, introversion rather by a negative. The extravert follows in his adjustment and response patterns more to the external, collectively valid norms, the ideals of the time, etc. the introvert's reaction, on the contrary, is mainly determined by subjective factors.

Unsuccessful Adjustment

Thence comes his so often unsuccessful adjustment to the external world. The extravert thinks feels and acts in reference to the object; he displaces his interest from the

ART FOR PSYCHOTHERAPY

subject out upon the object, he orientates himself predominantly by what lies outside him.

Orientation of Value

With the introvert the subject is the starting point of his orientation and the object is accorded at most a secondary, indirect value. This type of man draws back in the first moment in a given situation, as if with an unvoiced 'No', and only then follows his real reaction.

Empirical Material

Whereas the functional type describes the way in which the empirical material is specifically grasped and formed, the attitudinal type introversion-extraversion characterises the general psychological orientation, i.e., the direction of that general psychological energy which Jung conceives the libido to be.

It is anchored in our biological constitution and is much more firmly determined from birth than is our functional type.

Conscious Effort

For, although, the choice of the superior or principal function is in general determined by a certain constitutional predisposition to the differentiation of a particular function, this latter can be greatly modified by conscious effort or thought or even repressed.

ART FOR PSYCHOTHERAPY

Inner Rebuilding

This is very seldom the case with a basic attitude or manner of reaction. Here only an 'inner rebuilding', an alteration in the psyche's structure, can bring about such a change, either through a spontaneous transformation (in this case again biologically determined) in puberty or the climacteric years or through a toilsome process of psychic development such as an analysis.

Ancillary Functions

Therefore the differentiation of a second and third function, i.e., of the two ancillary functions, is relatively easier than that of the fourth, inferior function, for the latter is not only the furthest removed from the principal function and standing in sharpest contrast to it, but it also coincides with the still unlived, obscure attitudinal type. The introversion of the extraverted thinking type, for example, has not the tone of intuition or sensation but primarily that of feeling.

Compensatory Relation

Extraversion and introversion stand likewise in compensatory relation to each other. If consciousness is extraverts, the unconscious is introverted, and conversely. This fact is of decisive significance for psychological understanding.

Matrimonial Problems

The difference in types is thus the real psychological basis of matrimonial problems, difficulties between

ART FOR PSYCHOTHERAPY

children and parents, friction in relations of friendship and business, even indeed of social and political differences. Everything of which one is unconscious in one's own psyche appears in such cases projected upon the object, and as long as one does not recognise the projected content in one's own self the object is made into a scapegoat, the ethical task would then be to recognise in one's self the opposed attitudinal habitus, which is structurally given in everyone.

Eight Different Psychological Types

Combining extraversion and introversion as general attitudinal habitus with the four functions, these result in all eight different psychological types:

- the extraverted thinking type, the introverted thinking type,
- the extraverted feeling type, the introverted feeling type, etc.;

Compass

These (the Eight Different Psychological Types) form a kind of compass, with which we can orientate ourselves concerning the structure of the psyche.

ART FOR PSYCHOTHERAPY

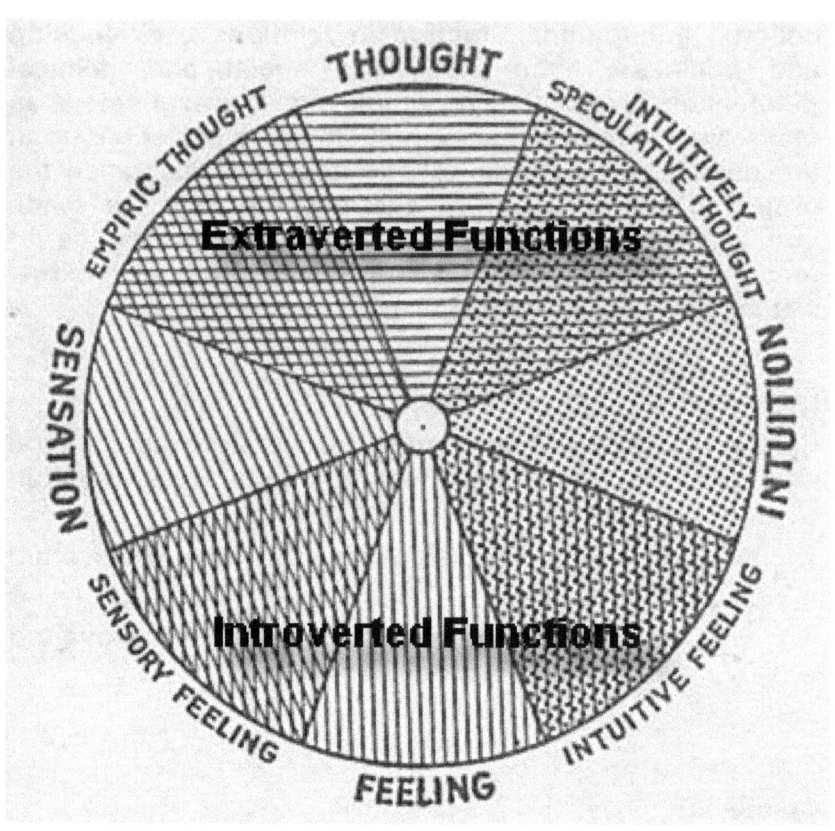

ART FOR PSYCHOTHERAPY

Sphere of Unconscious

As already mentioned, according to Jung's writings, the unconscious includes two regions, a personal, and a collective. The next diagram gives a schematic representation of this classification.

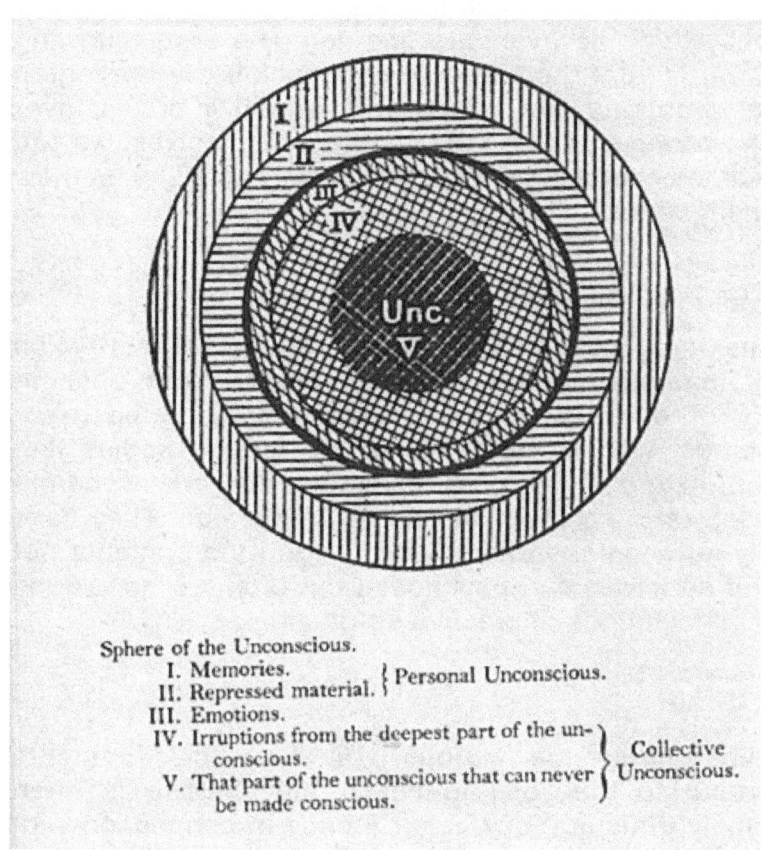

Sphere of the Unconscious.
  I. Memories.
  II. Repressed material. } Personal Unconscious.
  III. Emotions.
  IV. Irruptions from the deepest part of the unconscious.
  V. That part of the unconscious that can never be made conscious. } Collective Unconscious.

ART FOR PSYCHOTHERAPY

It has already been said what forms the content of the personal unconscious: forgotten, repressed, subliminally perceived thought, and felt matter of any kind.

Regional Divisions

But the collective unconscious is also divided into regions which lie over one another. The first, following downwards after the personal unconscious, is the region of our emotions and affects, the primitive drives, over which, however, when they manifest themselves, we can sometimes exercise control, which we can still somehow rationally order.

Autonomous Unconscious

The next region already includes those contents which break immediately out of the deepest, most obscure centre of our unconscious, never wholly to be made conscious, with elemental force, as foreign bodies that remain externally incomprehensible and never permit themselves to be assimilated fully by the ego. They have wholly autonomous character and form the contents not only of neuroses and psychoses but often of the visions and hallucinations of creative spirits.

Zones

To differentiate the various zones or their contents according to the zone to which they belong is often extremely difficult. They occur mostly in connection with each other, in a kind of mixture.

ART FOR PSYCHOTHERAPY

Symptom and Complex

The manifestations that first of all remain visible on the plane of consciousness are the symptom and the complex. The symptom can be defined as a phenomenon of the obstruction of the normal flow of energy and can manifest itself psychically or physically.

Broadening of Consciousness

It is a danger signal indicating that something essential in the conscious adjustment is disarranged or inadequate and that, accordingly, a broadening of consciousness ought to take place, i.e., a removal of the obstruction, although one is not always able to say in advance where the point of obstruction lies and how it is to be reached.

Arbitrary Functioning

Complexes Jung defines as "psychological parts split off from the personality, groups of psychic contents isolated from consciousness, functioning arbitrarily and autonomously, leading thus a life of their own in the dark sphere of the unconscious, whence they can at every moment hinder or further conscious acts."

Nuclear Element

The complex consists primarily of the 'nuclear element', which is mostly unconscious and autonomous and so beyond human influence, and secondarily of the numerous associations thereto, which in turn depend partly on the original personal disposition and partly

ART FOR PSYCHOTHERAPY

upon experiences casually connected with the environment.

Ascending Complex

The following diagram shows the ascending complex, under whose thrust consciousness, as it were, is broken through and the unconscious, lifting itself over the threshold of consciousness, forces itself onto the conscious plane.

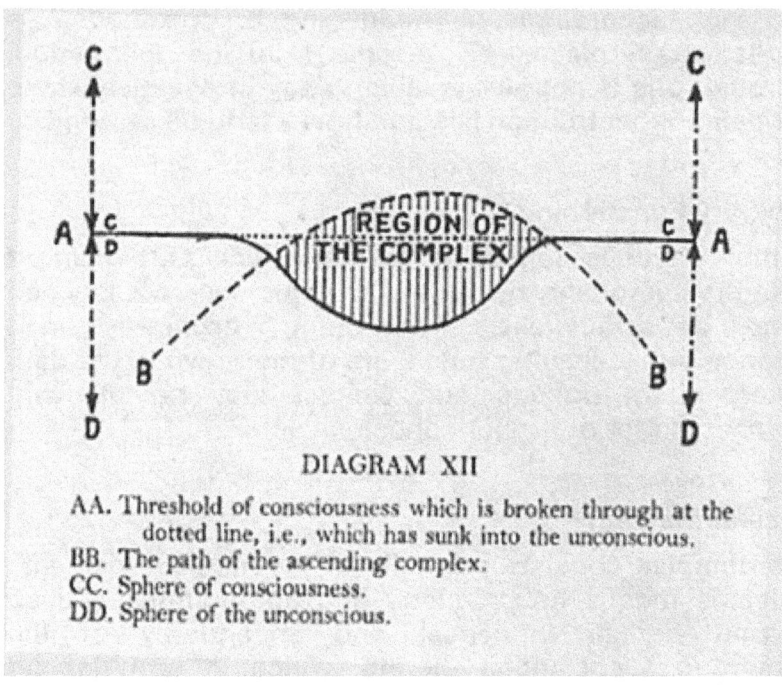

DIAGRAM XII

AA. Threshold of consciousness which is broken through at the dotted line, i.e., which has sunk into the unconscious.
BB. The path of the ascending complex.
CC. Sphere of consciousness.
DD. Sphere of the unconscious.

ART FOR PSYCHOTHERAPY

Passive State

The individual falls from an active, conscious state into a passive, 'possessed' one. Such an ascending complex acts as a foreign body in the field of consciousness. It has its specific closed-ness, wholeness, and relatively high degree of autonomy.

Moral Conflict

It generally gives the picture of a disordered psychic situation, strongly toned emotionally and incompatible with the habitual conscious situation or attitude. One of its most frequent causes is, accordingly, moral conflict; by no means limited to the sexual. The conflict is a mental power before which at times the conscious will and the freedom of the ego cease.

Complexes

Everyone has complexes. All sorts of everyday slips, as Freud in his Psychopathology of Everyday Life has shown, testify to that unmistakeably. Complexes do not necessarily imply inferiority of the individual who has them; they merely indicate that something un-united, inassimilable, conflicting exists, perhaps a hindrance, may be too a stimulus to greater efforts and so even to fresh successes.

ART FOR PSYCHOTHERAPY

Psychic Points

Complexes are thus in this sense focal and nodal points of psychic life with which one would not wish to dispense, indeed on could not do without, for else psychic activity would come to a standstill.

Emotional Shock

The origin of the complex is frequently in a so-called trauma, an emotional shock, or the like, by which a fragment of the psyche is split-off. The complex probably has its ultimate basis as a rule, however, in the apparent impossibility of accepting the whole of one's own individual nature.

Association Method

The actual significance of a complex can only be demonstrated and the freeing of the individual from its influence accomplished by practical psychotherapy. Its presence, its effective depth, and its emotional tone can nevertheless be determined with the aid of the association method worked out by Jung, based on his researching methods.

Psyche Mechanism

The association method has proven that the psyche mechanism is able to point with clock-wise exactness to complex-laden points of the psyche. Jung has worked out and refined the association method to the utmost precision, in manifold detail, and from the most different points of view.

ART FOR PSYCHOTHERAPY

Didactic and Diagnostic Method

As a didactic and diagnostic method it has become an essential aid to psychotherapy and belongs today to the standard equipment of psychiatric institutions, clinical psychological training, and vocational guidance of every kind, and even finds its use in the law courts. The concept of the complex comes from Jung.

Dream

The easiest and most effective way of acquainting one's self with the mechanisms and contents of the unconscious is via the dream, whose material consists of conscious and unconscious, familiar and unfamiliar elements. These elements can occur in the most varied mixtures and can be derived from anywhere, beginning with the so-called 'remnants of the day' and going on to the deepest contents of the unconscious.

Interpretation

Jung described their arrangement in the dream as standing outside of causality. Likewise space and time do not hold for them. The dream language is archaic, symbolic, pre-logical; a picture language whose meaning can only be discovered through special methods of interpretation.

Royal Pathway

Jung accords the dream extraordinary importance, regarding it not only as the way (the royal pathway) to the

ART FOR PSYCHOTHERAPY

unconscious but as a function through which in great part the unconscious exhibits its regulative activity. For the dream gives expression to the other side, the one opposite to the conscious attitude. Unswayable by our consciousness, it is a pure manifestation of the unconscious, of that uninfluenced primal nature that Jung on this account calls the objective psychic.

Continuity of Processes

Consciousness aims always at the adjustment of the individual to the external world. The unconscious, on the contrary, is indifferent to this egocentric purposive-ness and partakes of the impersonal objectivity of nature, whose one goal is maintenance of continuity of the psychic processes; it is accordingly a guard against one-sidedness, which could lead to isolation, inhibition, or other pathogenic phenomena.

Standard Symbols

In view of the already mentioned highly significant compensatory function of the dream, which not only expresses fears and wishes but profoundly affects the whole psychic situation, Jung refused to set up 'standard symbols'.

Manifold Contents

The contents of the unconscious are always manifold in meaning, and their significance depends equally upon the contents in which they occur and upon the specific external and internal situation of the dreamer. Many

ART FOR PSYCHOTHERAPY

dreams even go beyond the personal problems of the individual dreamer and are the expression of problems that occur over and over again in human history and concern the whole human collective.

Prophetic

In this case, Jung professes that these types of dreams have prophetic character and are therefore regarded even today among primitives as the concern of the entire tribe and are publicly interpreted with great ceremony.

Fantasies and Visions

Besides dreams, Jung distinguishes also fantasies and visions as bearers of the manifestations of the unconscious. They are related to dreams and occur in states of diminished consciousness. They exhibit a manifest and a latent content, are derived from the personal or collective unconscious, and furnish thus material equivalent to that of the dream for psychological interpretation. From the ordinary wish fulfilling dream to the ecstatic vision, pregnant with meaning, their variability is unlimited.

Universal Human History

Themes of mythological nature, whose symbolism illustrates universal human history, and reactions of particularly intensive kind, allow one to surmise the involvement of the deepest layers.

ART FOR PSYCHOTHERAPY

Archetypes

These motives and symbols Jung named Archetypes. They are representations of instinctive, i.e., psychologically necessary responses to certain situations, which, circumventing consciousness, lead by virtue of their innate potentialities to behaviour corresponding to the psychological necessity, even though it may not always appear appropriate when rationally viewed from without.

Conscious Adjustment

It is this absolute inner order of the unconscious that forms our refuge and help in the accidents and commotions of life, if we only understand how to get in touch with it. So it becomes comprehensible that the archetype can alter our conscious adjustment or even transform it.

Augustine's Term

The archetypes, Jung has borrowed this term from Augustine, are akin to what Plato called the 'idea'. Plato's idea may be understood only as a primordial image of highest perfection in its light aspect, aloof of earthly reality, whereas its dark counterpart does not belong to the world of eternity but to the ephemeral world of mankind.

Bipolar Structure

On the other hand, according to Jung's conception, the archetype is inherent in its bipolar structure; the dark

ART FOR PSYCHOTHERAPY

side as well as the light. Jung also calls the archetypes the 'organs of the soul'. They are only formally determined, not in regard to their contents; and their ultimate core of meaning may be delimited but never described.

Gestalt

If we wanted to look for further likely analogies the 'Gestalt' in the broadest sense of this term, as used in Gestalt psychology and also in biology, should be mentioned in the first place.

Axial System

"The form of these archetypes," Jung said, "is perhaps comparable to the axial system of a crystal, which predetermines the crystalline formation in the saturated solution, without itself possessing a material existence. This existence first manifests itself in the way the ions and then the molecules arrange themselves…"

Clarification

The diagram below attempts to clarify what has been stated by the Jungian followers and predominantly Jung himself.

ART FOR PSYCHOTHERAPY

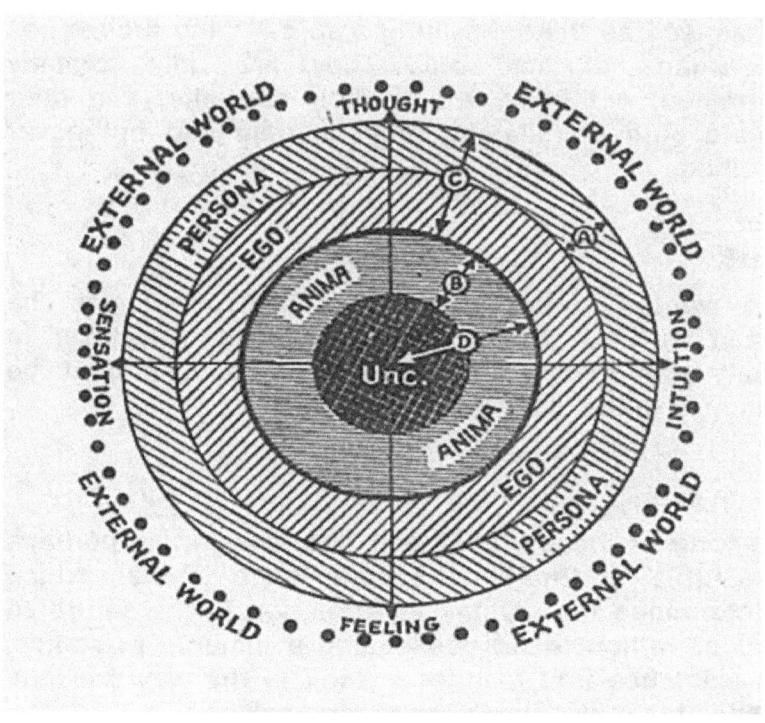

A. Would be the persona, lying as mediator between the ego and the external world,

B. The anima animus, represented as mediating function between the ego and the internal unconscious world.

C. Is at once ego and persona, representing the phenotypic, externally visible, manifest mental character,

D. Is the genotypic constituent, making up our invisible, latent, unconscious character.

ART FOR PSYCHOTHERAPY

Heavy Burden

The first plate shows the conjunction as it generally is but as it should not be. In the world of drives man and woman are here indivisibly one. Instead of striving together towards the sun of the spirit, they turn away from each other and sorrowfully carry it on backs as a heavy burden:

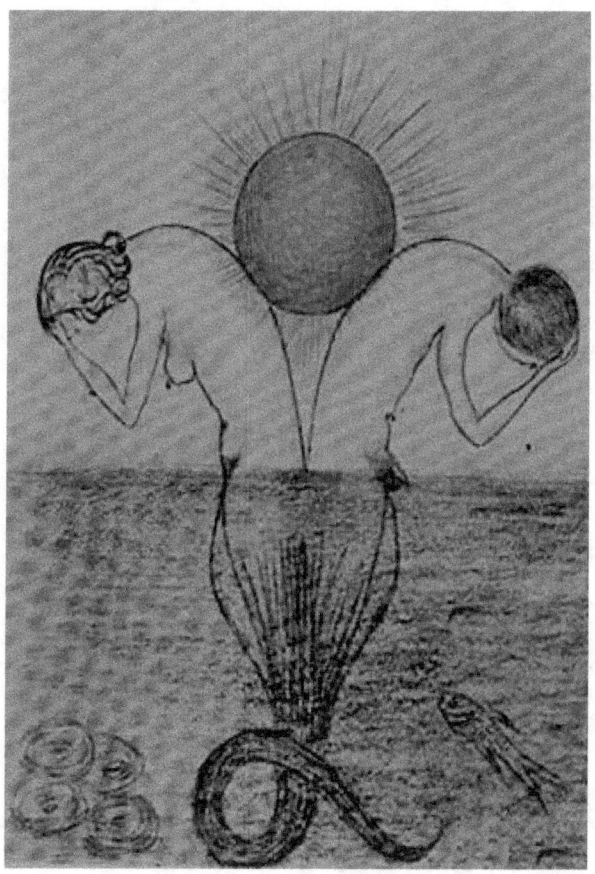

ART FOR PSYCHOTHERAPY

Creative Union

The next plate on the contrary shows the conjunction as a true and creative union. The unconscious animal sides of man and woman are not grown together indivisibly but join one another in the symbol of the 'healing snake', which helps them to raise out of the depths of the sea the symbol of the unconscious, the 'precious stone', the symbol of the Self, without which the interlacing Tree of Life could never blossom.

ART FOR PSYCHOTHERAPY

Psychic System

The drawing is intended to show that the Self both forms the centre and includes the surrounds the whole psychic system with power of its radiation. The different parts of the total psyche are likewise included in the diagram, without any claim being made to represent their order, positional value, etc., it being impossible to show anything so abstract schematically. The only content of the Self that we really know is the ego.

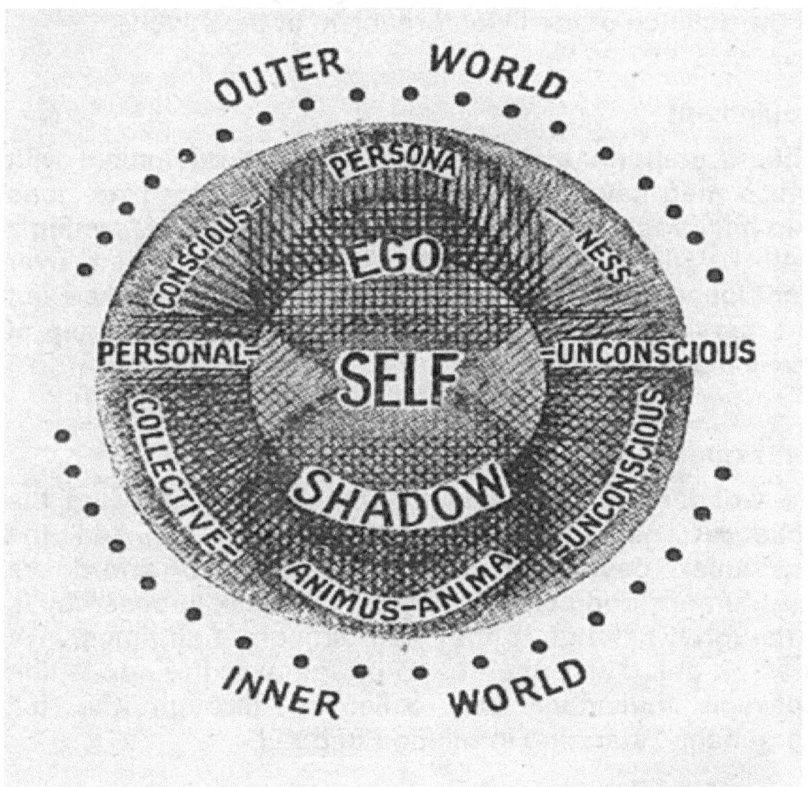

ART FOR PSYCHOTHERAPY

Jungian System: Neither Religion nor Philosophy

The Jungian system of psychotherapy claims, in spite of its intimate reference to the fundamental problems of our being, to be neither philosophy, nor religion. Jung's followers insist that the 'system' is the scientific summary and representation of all that the experienceable totality of the psyche includes; and as biology is the science of the living physical organism, so is the science of the living organism of the psyche.

Equipment

Thus it comprises also the whole of the equipment with which men have ever created and experienced religions and philosophies. It alone give the possibility of forming a well balanced living; that is not merely taken over traditionally and uncritically but that can be worked out and personally shaped by the individual with the help of these material and tools.

Reassurance and Comfort

No wonder that this system precisely today, when the collective psyche threatens to become all and the individual psyche nothing, is able to afford us reassurance and comfort; and that the task imposed by it, although it belongs to the most difficult of all times, lays it as an obligation upon us to bridge over the opposition between individual and collective through the full personality, standing in relation to both!

ART FOR PSYCHOTHERAPY

Reason over Instinctive Nature

The predominance which our reason, our one-sidedly differentiated intellect, has gained in the West over our instinctive nature and which expresses itself in our highly developed civilisation in a masterful technic that seems to have lost every connection with the external depth of the psyche, can only be compensated by calling to aid the creative powers lying there, restoring them to their rights, and elevating them to the heights of this intellect.

Transformation

This transformation, however, can only begin with the individual, for the masses are blind beasts. If this transformed individual has recognised himself as God's likeness in the deepest ethical sense of obligation, then (as Jung says) "he will be on the one hand excellent in knowledge, on the other excellent in will, and no arrogant superman!"

Culture of the Future

As a finale to this précis on the science of Jungian psychotherapy one can maintain that the responsibility and the task of the culture of our future belong more than ever to the individual!

Psychotherapies

Psychotherapy implies the treatment of mental discomfort, dysfunction, or disease by psychological means by a trained therapist who adheres to a particular theory of both symptom causation and symptom relief.

ART FOR PSYCHOTHERAPY

The American psychotherapist Jerome Frank has classified psychotherapies into religio-magical and empirico-scientific forms. The former depend on the shared beliefs of the therapist and client in magic, spirits, or other supernatural processes or powers. This article is concerned, however, with the latter forms of psychotherapy, which have been developed by modern medicine and which are carried out by a member of one of the mental health professions such as a psychiatrist or a clinical psychologist.

It is usual to contrast two main forms of psychotherapy, dynamic and behavioural. They are conceptually different; behaviour therapy concentrates on alleviating a patient's overt symptoms, which are attributed to faulty learning, while dynamic therapy concentrates on understanding the meaning of symptoms and understanding the emotional conflicts within the patient that may be causing those symptoms. In their pure forms the two approaches are very different, but in practice many therapists use elements of both.

Psychiatry

Psychiatry is the branch of medicine that is concerned with the diagnosis, treatment, and prevention of mental disorders.

The term psychiatry is derived from two Greek words meaning "mind healing." Until the 18th century, mental illness or disorder was most often seen as demonic possession, but it gradually came to be considered as a sickness requiring treatment.

ART FOR PSYCHOTHERAPY

Many judge that modern psychiatry was born with the efforts of Philippe Pinel in France and J. Connolly in England, who both advocated a more humane approach to mental illness. By the 19th century, research, classification, and treatment of disorders had gained momentum. Psychotherapy evolved from its origins in spiritual healing.

The psychoanalytic theory of Sigmund Freud and his followers dominated the field for many years and did not receive a serious theoretical challenge until behaviour therapy and therapies deriving from humanistic psychology were developed in the mid-20th century. Insight therapies such as psychoanalysis, which pursue greater awareness of the patient's internal conflicts, continue to be dominant in psychiatric practice.

The trained psychiatrist, who has completed medical school and a psychiatric residency, commonly employs medical treatments in addition to psychotherapy. Lobotomy, or leucotomy, whereby nerve fibres running to the front of the brain are severed, is today used only in severe cases and has generally lost favour as a treatment.

Shock therapy (also called electroshock, or electroconvulsive, therapy) continues to be used for severe depressions and certain forms of psychosis. The medical technique that is by far the most widely used is drug therapy. The advent in the 1950s of psychotropic (mind-altering) drugs revolutionized treatment of the mental patient. Like the other medical techniques, drug therapy has sometimes been abused in pursuit of patient "management"; used properly, however, it can enhance a

ART FOR PSYCHOTHERAPY

patient's outlook for recovery and return to the community.

The contemporary psychiatrist frequently functions as a member of a mental-health team that includes clinical psychologists and social workers. As the therapeutic roles of these three professionals are not necessarily clearly delineated, an uneasy balance in orientation and division of skills may exist.

Child Development

This term refers to the growth of perceptual, emotional, intellectual, and behavioural capabilities and functioning during childhood. The term childhood denotes that period in the human lifespan from the acquisition of language at one or two years to the onset of adolescence at 12 or 13 years.

The end of infancy and the onset of childhood are marked by the emergence of speech at one to two years of age. Children make enormous progress in language acquisition in their second year and demonstrate a continually growing vocabulary, an increasing use of words in combinations, and a dawning understanding of the rules of grammar and syntax. By their third year children tend to use sentences containing five or even six words, and by the fourth year they can converse in adult-like sentences. Five- and six-year-olds demonstrate a mastery of complex rules of grammar and meaning.

Early childhood (two to seven years) is also the time in which children learn to use symbolic thought and language to manipulate their environment. They learn to perform various mental operations using symbols,

ART FOR PSYCHOTHERAPY

concepts, and ideas to transform information they gather about the world around them. The beginnings of logic, involving the classification of ideas and an understanding of time and number, emerge in later childhood (7 to 12 years).

Children's memory capacity also advances continually during childhood and underpins many other cognitive advances they make at that time. As both short-term and long-term memory improves, children demonstrate an increasing speed of recall and can search their memory for information more quickly and efficiently.

Young children's growing awareness of their own emotional states, characteristics, and abilities lead to empathy--i.e., the ability to appreciate the feelings and perspectives of others. Empathy and other forms of social awareness are in turn important in the development of a moral sense. The basis of morality in children may be said to progress from a simple fear of punishment and pain to a concern for maintaining the approval of one's parents.

Another important aspect of children's emotional development is the formation of their self-concept, or identity--i.e., their sense of who they are and what their relation to other people is. Sex-role identity, based on gender, is probably the most important category of self-awareness and usually appears by the age of three.

The onset of the physical and emotional changes of puberty and the acquisition of the logical processes of adults mark the end of childhood and the start of adolescence.

ART FOR PSYCHOTHERAPY

Child Psychology

Child psychology is the study of the psychological processes of children, specifically, how these processes differ from those of adults, how they develop from birth to the end of adolescence, and how and why they differ from one child to the next. The topic is sometimes subsumed with infancy, adulthood, and aging under the category of developmental psychology.

As a scientific discipline with a firm empirical basis, child study is of comparatively recent origin. It was initiated in 1840, when Charles Darwin began a record of the growth and development of one of his own children, collecting the data much as if he were studying some strange species. A similar, more elaborate study was published by the German psycho-physiologist W.T. Preyer (Die Seele des Kindes [1882; The Mind of the Child]) that set the fashion for a series of others. In 1891 the American educational psychologist G. Stanley Hall established a periodical, the Pedagogical Seminary, devoted to child psychology and pedagogy. During the early 20th century, the development of intelligence tests and the establishment of child guidance clinics further defined the field of child psychology.

A number of notable 20th-century psychologists--among them Sigmund Freud, Melanie Klein, and Freud's daughter, Anna Freud--dealt with child development chiefly from the psychoanalytic point of view. Perhaps the greatest direct influence on modern child psychology was Jean Piaget of Switzerland. By means of direct observation and interaction, Piaget developed a theory based on the systematic study of the acquisition of understanding in children. He described the various

ART FOR PSYCHOTHERAPY

stages of learning in childhood and characterized the child's perception of himself and the world at each stage.

The data of child psychology are gathered from a variety of sources. Observations by relatives, teachers, and other adults, as well as the psychologist's direct observation of and interviews with a child (or children), provide a significant amount of material. In some cases a one-way window or mirror is used so that children are free to interact with their environment or others without awareness that they are being watched. Projective tests, personality and intelligence tests, and experimental methods have also proved useful in understanding child development.

Despite attempts to unify the various theories of child development, the field remains dynamic, developing as human understanding of physiology and psychology changes.

Moral Standards

Through the ages man has been pre-occupied with moral standards, probably more than other philosophical concepts. Societies through their various stages of evolution varied the theme with distinct differences in their demands on standards. These codes of behaviour were influenced by religions and the dogmas of each regional culture. What is acceptable in one tribe may be a fatal error in another society.

These variations of morality place a big demand on therapists. Individual clients from different cultures, social groups, of known and unknown social norms may occupy the practitioner's couch; the significance of this

ART FOR PSYCHOTHERAPY

being the understanding demanded of the therapist. There is no doubt that to establish the appropriate method for the agreed upon treatment is a time consuming exercise, for both parties.

People live in groups and humans choose to live in states, simply for what they can get out of society. Those who choose to live in solitude become recluse in monasteries and nunneries, or become thinkers in isolation high up in mountains. The remaining who wish to explore their interaction with others, integrate within the moral and legal demands of a government.

ART FOR PSYCHOTHERAPY

SECTION SIX: CINEMATOGRAPHY AND TECHNOLOGY

Motion pictures

This is the means for the production and showing of motion pictures. It includes not only the motion-picture camera and projector but also such technologies as those involved in recording sound, in editing both picture and sound, in creating special effects, and in producing animation.

Motion-picture technology is a curious blend of the old and the new. In one piece of equipment state-of-the-art digital electronics may be working in tandem with a mechanical system invented in 1895. Furthermore, the technology of motion pictures is based not only on the prior invention of still photography but also on a combination of several more or less independent technologies; that is, camera and projector design, film manufacture and processing, sound recording and reproduction, and lighting and light measurement.

History

Motion-picture photography is based on the phenomenon that the human brain will perceive an illusion of continuous movement from a succession of still images exposed at a rate above 15 frames per second. Although posed sequential pictures had been taken as early as 1860, successive photography of actual movement was not achieved until 1877, when Eadweard Muybridge used 12 equally spaced cameras to demonstrate that at some time all four hooves of a galloping horse left the ground at once. In 1877–78 an associate of Muybridge devised a

ART FOR PSYCHOTHERAPY

system of magnetic releases to trigger an expanded battery of 24 cameras.

The Muybridge pictures were widely published in still form. They were also made up as strips for the popular parlour toy the zoetrope "wheel of life," a rotating drum that induced an illusion of movement from drawn or painted pictures. Meanwhile, Émile Reynaud in France was projecting sequences of drawn pictures onto a screen using his Praxinoscope, in which revolving mirrors and an oil-lamp "magic lantern" were applied to a zoetrope-like drum, and by 1880 Muybridge was similarly projecting enlarged, illuminated views of his motion photographs using the Zoöpraxiscope, an adaptation of the zoetrope.

Although a contemporary observer of Muybridge's demonstration claimed to have seen "living, moving animals," such devices lacked several essentials of true motion pictures. The first was a mechanism to enable sequence photographs to be taken within a single camera at regular, rapid intervals, and the second was a medium capable of storing images for more than the second or so of movement possible from drums, wheels, or disks.

A motion-picture camera must be able to advance the medium rapidly enough to permit at least 16 separate exposures per second as well as bring each frame to a full stop to record a sharp image. The principal technology that creates this intermittent movement is the Geneva watch movement, in which a four-slotted star wheel, or "Maltese cross", converts the tension of the mainspring to the ticking of toothed gears. In 1882 Étienne-Jules Marey employed a similar "clockwork train"

ART FOR PSYCHOTHERAPY

intermittent movement in a photographic "gun" used to "shoot" birds in flight.

Twelve shots per second could be recorded onto a circular glass plate. Marey subsequently increased the frame rate, although for no more than about 30 images, and employed strips of sensitized paper (1887) and paper-backed celluloid (1889) instead of the fragile, bulky glass. The transparent material trade-named celluloid was first manufactured commercially in 1872.

It was derived from collodion, that is, nitrocellulose (gun cotton) dissolved in alcohol and dried. John Carbutt manufactured the first commercially successful celluloid photographic film in 1888, but it was too stiff for convenient use.

By 1889 the George Eastman company had developed a roll film of celluloid coated with photographic emulsion for use in its Kodak still camera. This sturdy, flexible medium could transport a rapid succession of numerous images and was eventually adapted for motion pictures.

Thomas Edison is often credited with the invention of the motion picture in 1889. The claim is disputable, however, specifically because Edison's motion-picture operations were entrusted to an assistant, W.K.L. Dickson, and generally because there are several plausible pre-Edison claimants in England and France.

Indeed, a U.S. Supreme Court decision of 1902 concluded that Edison had not invented the motion picture but had only combined the discoveries of others. His systems are important, nevertheless, because they prevailed commercially.

ART FOR PSYCHOTHERAPY

The heart of Edison's patent claim was the intermittent movement provided by a Maltese cross synchronized with a shutter. The October 1892 version of Edison's Kinetograph camera employed the format essentially still in use today. The film, made by Eastman according to Edison's specifications, was 35 millimetres (mm) in width. Two rows of sprocket holes, each with four holes per frame, ran the length of the film and were used to advance it. The image was 1 inch wide by $3/4$ inch high.

At first Edison's motion pictures were not projected. One viewer at a time could watch a film by looking through the eyepiece of a peep-show cabinet known as the Kinetoscope. This device was mechanically derived from the zoetrope in that the film was advanced by continuous movement, and action was "stopped" by a very brief exposure. In the zoetrope, a slit opposite the picture produced a stroboscopic effect; in the Kinetoscope the film traveled at the rate of 40 frames per second, and a slit in a 10-inch-diameter rotating shutter wheel afforded an exposure of 6,000 second. Illumination was provided by an electric bulb positioned directly beneath the film.

The film ran over spools. Its ends were spliced together to form a continuous loop, which was initially 25 to 30 feet long but later was lengthened to almost 50 feet. A direct-current motor powered by an Edison storage battery moved the film at a uniform rate.

The Kinetoscope launched the motion-picture industry, but its technical limitations made it unsuitable for projection. Films may run continuously when a great deal of light is not crucial, but a bright, enlarged picture requires that each frame be arrested and exposed intermittently as in the camera. The adaptation of the

ART FOR PSYCHOTHERAPY

camera mechanism to projection seems obvious in retrospect but was frustrated in the United States by Dickson's establishment of a frame rate well above that necessary for the perception of continuous motion.

After the Kinetoscope was introduced in Paris, Auguste and Louis Lumière produced a combination camera/projector, first demonstrated publicly in 1895 and called the *cinématographe*. The device used a triangular "eccentric" (intermittent) movement connected to a claw to engage the sprocket holes.

As the film was stationary in the aperture for two-thirds of each cycle, the speed of 16 frames per second allowed an exposure of $1/25$ second. At this slower rate audiences could actually see the shutter blade crossing the screen, producing a "flicker" that had been absent from Edison's pictures. On the other hand, the hand-cranked *cinématographe* weighed less than 20 pounds (Edison's camera weighed 100 times as much).

The Lumière units could therefore travel the world to shoot and screen their footage. The first American projectors employing intermittent movement were devised by Thomas Armat in 1895 with a Pitman arm or "beater" movement taken from a French camera of 1893. The following year Armat agreed to allow Edison to produce the projectors in quantity and to market them as Edison Vitascopes. In 1897 Armat patented the first projector with four-slot star and cam (as in the Edison camera).

One limitation of early motion-picture filming was the tearing of sprocket holes. The eventual solution to this problem was the addition to the film path of a slack-forming loop that restrained the inertia of the take-up reel.

ART FOR PSYCHOTHERAPY

When this so-called Latham loop was applied to cameras and projectors with intermittent movement, the growth and shrinkage of the loops on either side of the shutter adjusted for the disparity between the stop-and-go motion at the aperture and the continuous movement of the reels.

When the art of projection was established, the importance of a bright screen picture was appreciated. Illumination was provided by carbon arc lamps, although flasks of ether and sticks of unslaked calcium ("limelight") were used for brief runs.

Introduction of sound

The popularity of the motion picture inspired many inventors to seek a method of reproducing accompanying sound. Two processes were involved: recording and reproducing. Further, the sound reproduction had to be presented in an auditorium and had to be quite good.

This could not be achieved without a good amplifier of electrical signals. In 1907 Lee De Forest invented the Audion, a three-element vacuum tube, which provided the basis in the early 1920s for a feasible amplifier that produced an undistorted sound of sufficient loudness.

Next came the problem of synchronization of the sound with the picture. A major difficulty turned out to be the securing of constant speed in both the recorder and reproducer.

Many ingenious ideas were tried. In 1918 in Germany, the use of a modulated glow lamp in photographically recording sound and a photocell for reproduction were studied. In Denmark in 1923, an oscillograph light

ART FOR PSYCHOTHERAPY

modulator and selenium-cell reproducer were developed. De Forest tried a gas-filled glow discharge operated by a telephone transmitter to record a synchronized sound track on the film. For loudspeakers he experimented with a variety of devices but finally chose the speaker with horn.

The operating signal was obtained from a light shining through the film sound track and detected by a light-sensitive device (photocell). These were used in a system called Phonofilm, which was tried experimentally in a number of theatres. In 1927 the Fox Film Corporation utilized some of these principles in the showing of Fox Movietone News.

Meanwhile, the Western Electric Company laboratories in the United States had been making extensive studies on the nature of speech and other sounds and on techniques for recording and reproducing such sounds.

They experimented with recording on a phonograph disc and developed a 16-inch (40.6-centimetre) disc rotated at $33\ 1/3$ revolutions per minute; they improved loudspeakers, introduced the moving-coil type of speaker, and generally improved the entire electronic amplification system. The Warner Bros. movie studio became interested in all these developments and formed the Vitaphone Corporation to market the complete system.

Warner Bros. premiered Vitaphone in 1926 with a program featuring short musical performances and a full-length picture, *Don Juan*, which had synchronized music and effects but no speech. In 1927 it brought out *The Jazz Singer*, which was essentially a silent picture with Vitaphone score and sporadic episodes of synchronized

ART FOR PSYCHOTHERAPY

singing and speech. Warners presented the first "100-percent talkie," *The Lights of New York*, in 1928.

Although the Vitaphone system offered fidelity superior to sound-on-film systems at this stage, it became clear that recording on film would be much more convenient. Among other disadvantages, it was extremely difficult with the wax discs to shoot outdoors or to edit sound. By 1931 Warner Bros. ceased production of sound-on-disc and adopted the sound-on-film option preferred by the other studios.

Sound-on-film, a system that in various guises had enjoyed several periods of popularity, underwent constant improvements in the 1910s and 1920s. Although a sound track on the picture negative was used for Movietone News,

Fox's dramatic productions used a separate sound film on fine-grain print stock that could be edited apart from the picture yet in synchronism with it. One serious problem of sound-on-film systems had been the distortion of the signal introduced by the glow lamp when recording the sound track on film.

The Western Electric Company devised a "double-string" light valve. A wire was looped around a post and parallel to itself. When speech current was applied to the wire in a magnetic field, the wire vibrated toward and away from itself according to the applied electrical waveform. A steady beam of white light shining through the loop was modulated in intensity by the varying gap between the wires; the modulated beam was photographed while masked by a slit perpendicular to the edge of the film.

The resulting sound track appeared as darker or fainter parallel lines on the edge of the film. Known as the

ART FOR PSYCHOTHERAPY

variable density system, this method of optically recording sound was originally used by all but one of the major Hollywood studios.

The Radio-Keith-Orpheum Corporation (RKO) was created in 1928 to showcase the Radio Corporation of America (RCA) Photophone system of variable area recording. With this system, the sound recording was modulated by a rotating mirror and the slit was parallel to the edge of the film; reproduction employed the perpendicular slit of the variable density sound track.

Minor problems of incompatibility between recording and reproduction were solved in late 1928 when the track was narrowed down to stay safely within the area scanned by the beam. Identical side-by-side tracks were employed to compensate for lateral misalignment. Initially inferior in quality, the variable area system gradually drew even with the quality of the density system and supplanted it altogether in the 1950s.

Whereas there was wide variation in the speed at which silent films were photographed and projected, sound necessitated standardization of the frame rate. In 1927 the speed was standardized at 24 frames per second, or 90 feet per minute for 35-mm film.

The development of sound technology in the first years of talking pictures focused on two areas. One involved the development of blimped cameras, directional microphones, microphone booms, and quieter lights, so that sound could be recorded more cleanly at the time of shooting. The other technologies involved the ability to add, edit, and mix sound separately from the time the picture was recorded.

ART FOR PSYCHOTHERAPY

Introduction of colour

From their earliest days, silent films could be coloured using non-photographic methods. One means was to hand-colour frames individually. Another method made it possible to use monochrome sections for mood (*e.g.,* blue for night scenes or red for passionate sequences). Monochrome stock was created by "tinting" the film base or "toning" the emulsion (by bathing the film in chemical salts).

The photography of colour was theorized decades before it was developed for motion pictures. In 1855 the British physicist James Clerk Maxwell argued that a full-colour photographic record of a scene could be made by filming three separate black-and-white negatives through filters coloured, respectively, red, green, and blue, the three primary colours.

When converted to positives, the transparent exposed areas of the three films could pass light through the appropriate filter to produce three images, one red, one green, and one blue. Superimposing the three images would "rebuild" the image in its original colours.

In 1868 Louis Ducos du Hauron identified the additive and subtractive systems of colour. Both systems originate as red, green, and blue negative records. The difference occurs in the positive image, which may be composited from either the additive or subtractive primaries.

The subtractive primaries—cyan, magenta, and yellow—are the complements of the additive primaries and can be obtained by subtracting, respectively, red, green, and

ART FOR PSYCHOTHERAPY

blue from white. (Subtracting all three additive primaries yields black; adding all three yields white.)

In motion-picture prints, overlapping dye layers in the three subtractive primaries are simultaneously present on a clear, transparent base, and the image is projected with an exposure of white light. The dark areas of the cyan layer subtract all red colour, permitting only cyan (the mixture of blue and green) to pass through; the transparent areas pass all the white light. The magenta and yellow layers act similarly, and the original colour image is reproduced. The fineness of resolution is limited only by the structure of photographic grain or dye globules.

The first film colour systems were additive, but they were confronted by insurmountable limitations. In an additive system, the three colour records remain discrete and meet only as light rays on the screen. The best picture results when a separate film is made for each colour; however, each colour can occupy alternating frames or small, alternating portions of each frame of a single film.

(A contemporary example of additive colour can be seen in projection television, in which red, green, and blue lenses converge to produce an image so enlarged that the separate colour areas, or dots, become discernible.)

The best known of the early additive processes was Kinemacolor (1906), which, for manageability, reduced the three colour records to two: red-orange and blue-green. A single black-and-white film was photographed and projected at 32 frames per second (twice the normal silent speed) through a rotating colour filter.

ART FOR PSYCHOTHERAPY

The two colour records occupied alternate frames and were integrated by the retention characteristic of the human eye.

As there were no separate red-orange and blue-green records for each image, displacement from frame to frame was visible during rapid movement, so that a horse might appear to have two tails. Inventors tried to increase the film speed, reduce the frame size, or combine two films with mirrored prisms, but additive systems continued to be plagued by excessive film consumption, poor resolution, loss of light, and registration problems.

The first subtractive process employing a single film strip in an ordinary projector without filters was Prizma Color in 1919. (Prizma Color had been introduced as an additive process but was soon revised.) The basis was an ingenious "duplitized" film with emulsion on both sides. One side was toned red-orange and the other blue-green. The stock long outlasted the Prizma company and was in use as late as the early 1950s in such low-cost systems as Cinecolor.

Similar enough to provoke litigation was an early (1922) process by Technicolor in which separate red and green films were cemented back-to-back, resulting in a thick and stiff print that scratched easily. Although only four two-colour Technicolor features were produced by the end of the silent era, Technicolor sequences were a highlight of several big-budget pictures in the mid-to-late 1920s, including

The Phantom of the Opera (1923–25) and *Ben Hur* (1925). Technicolor devised the first of its dye-transfer, or imbibition, processes in 1928. Red and green dye images

ART FOR PSYCHOTHERAPY

were printed onto the same side of clear film containing a black silver sound track.

When Technicolor's appeal seemed on the wane, it devised a greatly improved three-register process (1932). The perfected Technicolor system used a prism/mirror beam-splitter behind a single lens to record the red, green, and blue components of each image on three strips of black-and-white film.

Approximately one-third of the light was transmitted to the film behind a green filter in direct path of the lens; the film was sensitized to green light by special dyes. A partially silvered mirror (initially flecked with gold) directed the remainder of the light through a magenta (red plus blue) filter to a bi-pack of orthochromatic and panchromatic films with their emulsion surfaces in contact. The orthochromatic film became the blue record.

As it was insensitive to red light, the orthochromatic film passed the red rays to the panchromatic film. A 1938 improvement added red-orange dye to the orthochromatic film so that only red light reached the panchromatic layer. In 1941 Monopack Technicolor was introduced. This was a three-layer film from which separation negatives were made for the Technicolor dye-transfer printing process.

Using the dye-transfer method, it was necessary to make gelatin positives that contained the image in relief. Dye filled the recesses while the higher areas remained dry. Each gelatin matrix thus imprinted its complement onto the film base.

As in the two-colour process, a black silver sound track was printed first on clear film. When magnetic sound became popular, the oxide strips were embossed after printing. Technicolor gave excellent results but was very expensive.

In 1936 Germany produced Agfacolor, a single-strip, three-layer negative film and accompanying print stock. After World War II Agfacolor appeared as Sovcolor in the Eastern bloc and as Anscocolor in the United States, where it was initially used for amateur filmmaking. The first serious rival to Technicolor was the single-strip Eastmancolor negative, which was introduced in 1952 by the Eastman Kodak Company but was often credited under a studio trademark (*e.g.,* Warnercolor).

Eastmancolor did not require special camera or processing equipment and was cheaper than Technicolor. Producers naturally preferred the less expensive Eastmancolor, especially since they had, in response to the perceived threat of television, increased production of colour films. (After the 1960s black-and-white films were so rare that they cost more to print than colour films.) The 1950s vogue for CinemaScope and three-dimensional productions, both incompatible with the Technicolor camera, also hastened the demise of Technicolor photography.

Dye-transfer printing remained cost-effective somewhat longer, but Technicolor was forced to abandon the process in the 1970s. This has created a significant problem for film preservationists because only Technicolor film permanently retains its original colours. Other colour prints fade to magenta within seven years, yet the hard gelatin dyes of a Technicolor print remain undimmed even after the film's nitrate base has begun to decompose.

ART FOR PSYCHOTHERAPY

In the 1980s computerized versions of the hand-stencilled colour films of the silent era were developed to rejuvenate old black-and-white films for video.

Wide-screen and stereoscopic pictures

Until the early 1950s, the screen shape, or aspect ratio (expressed as the ratio of frame width to frame height), was generally 1.33 to 1, or 4 to 3. In the mid-1950s the ratio became standardized at 1.85 to 1 in the United States and 1.66 or 1.75 to 1 in Europe. These slightly wider images were accomplished by using the same film but smaller aperture plates in the projector and by using shorter-focal-length lenses.

Many people have felt that, while vision at the extreme sides of the vision field does not usually contribute much information to the eyes, it does add substantially to the illusion of reality when it is present. Hence, there have been periods when film producers have attempted to introduce extremely wide formats. As early as 1929, Grandeur films were presented using 70-mm instead of the standard 35-mm film to give a wider field of view.

In 1952 a radical attack was made on wide-screen projection in the form of the Cinerama, which used three projectors and a curved screen. The expanded field of view gave a remarkable increase in the illusion of reality, especially with such exciting and spectacular subjects as a ride down a toboggan slide.

There were technical problems, including the necessity of carrying three cameras bolted together at the correct angles on the toboggan or other carrier, synchronization of the three separate films, and matching of the image structure and brightness at the joining edges on the screen. After 1963 Cinerama replaced its three-film process with a 70-mm anamorphic system with an aspect ratio of 2.75 to 1.

The use of anamorphic lenses for wide-screen projection was introduced by CinemaScope in 1953. An anamorphic optical system photographs with a different magnification horizontally than it does vertically. The lens seems to squeeze the image so that on the film itself figures appear tall and thin. A lens on the projector reverses the effect, so that the images on the screen reacquire normal proportions.

In 1955 Todd-AO introduced a wider film (photographed on a 65-mm negative and printed on a 70-mm positive for projection), with several sound tracks added. Like anamorphic systems, the wider format could be achieved with a single projector. The first two Todd-AO productions, *Oklahoma!* (1955) and *Around the World in 80 Days* (1956), were made at 30 frames per second for a nearly flicker-free image; 70-mm films are now photographed and projected at 24 frames per second.

Amusement parks and world's fairs have often featured 360-degree projection. The first system was presented at the Disneyland amusement park in 1955. At first, the projection involved 11 16-mm projectors and screens and, later, nine 35-mm projectors.

The audience stood on a low platform in the middle. The result was extremely realistic. In one scene, showing the view from a cable car in San Francisco, the viewers were seen involuntarily to lean over on the curves, as if they were actually on the cable car. The format, however, has limited uses for general storytelling.

ART FOR PSYCHOTHERAPY

In the 1980s, efforts to improve picture quality took two routes: increase in frame rate (Showscan operates at 60 frames per second) or increase in overall picture size—height as well as width (IMAX and Futurevision). In these formats the sound tracks are usually printed on a separate, magnetic strip of film.

Another project intended to improve the illusion of reality in motion pictures has been stereoscopic, or three-dimensional, cinematography. "3-D" films use two cameras or one camera with two lenses. The centres of the lenses are spaced $2\ 1/2$ to $2\ 3/4$ inches apart to replicate the displacement between a viewer's left and right eyes. Each lens records a slightly different view corresponding to the different view each eye sees in normal vision.

Despite many efforts to create "3-D without glasses" (notably in the U.S.S.R., where a screen of vertical slats was used for many years), audience members have had to wear one of two types of special glasses to watch 3-D films. In the early anaglyph system, one lens of the glasses was red and the other green (later blue).

The picture on the screen viewed without glasses appeared as two slightly displaced images, one with red lines, the other with green. Each lens of the glasses darkened its opposite colour so that each eye would see only the image intended for it.

The Polaroid system, used for commercial 3-D movies since the early 1950s, is based on a light-polarizing material developed by the American inventor Edwin H. Land in 1932. In this method, known as Natural Vision, two films are recorded with lenses that polarize light at different angles. The lenses on the glasses worn by spectators are similarly polarized so that each admits its corresponding view and blocks the other.

Early versions of Polaroid 3-D used two interlocked projectors to synchronize the two pictures. A later system, revived in the 1970s and 1980s, stacked the left and right components vertically on half-frames two sprocket holes high. The images were converged by means of a mirror and/or prism.

Professional motion-picture production

The principles of operation of modern professional motion-picture cameras are much the same as those of earlier times, although the mechanisms have been refined. A film is exposed behind a lens and is moved intermittently, with a shutter to stop the light while the film is moving. In the process, the film is unrolled from a supply reel, through the intermittent to the gate where the exposure takes place, and then on to the take-up reel.

Principal parts

Lenses have gone through a continuous evolution in the last half century, for both still and motion-picture photography. The two major objectives have been to focus properly all the colours of the image at the film plane (*i.e.,* to make the lens achromatic) and to focus portions of a beam coming from different portions of the lens, the centre or the edges, at the same point on the film (*i.e.,* anastigmatic).

Both objectives require solution for as large a lens opening as possible, in order to capture maximum light for the exposure, and for as wide a field of view as will be

ART FOR PSYCHOTHERAPY

needed in the use of the lens. In order to solve these problems, lenses have been made with more and more components.

Also, more types of glass have been discovered and developed, to give better achromatic performance. It was found, about 1939, that a special coating of the glass-to-air surface of a lens component could greatly diminish reflections from this surface without affecting other properties of the lens.

The use of such coatings improved image contrast by reducing the stray rays that were produced by reflections in a multiple-component lens. Coatings also reduce loss of light by reflection in the desired rays. Coating developments have permitted the manufacture of lenses with many more components than had previously been possible.

Long experience with both motion-picture and still cameras has shown the need for a variety of focal lengths (ranging from ultrawide angle to telephoto) to photograph scenes under the best conditions. To make changing focal lengths more convenient, the lenses have sometimes been mounted on a turret, so that one out of a set of three lenses may be quickly selected. For motion pictures this would mean an interruption in the action depicted. A continuous change would be more desirable.

When two lenses are used in a tandem combination, the focal length of the combination varies according to the separation between the two components. For example, when two thin converging lenses are mounted close together, the combined focal length is shorter than when they are separated a certain distance.

ART FOR PSYCHOTHERAPY

Thus, the focal length of the combination can be continuously varied over a range merely by changing the separation.

This observation led to the conception of camera lenses of variable focal length in which the variation is obtained by moving one or more elements. One simple design consists of two fixed convex (converging) lenses of unequal power with a movable concave (diverging) lens between them.

When the central concave lens is located close to the front convex element, the combination focal length can be shorter (and the image therefore smaller) than when it is located close to the rear convex element.

The design can be made such that, with the two convex elements remaining fixed, a distant view can remain almost in focus on the film as the middle element is moved. Exact focus for this arrangement, however, could not be attained.

Thus, for lenses of this design, a cam device has in the past been provided to move the front convex element a short distance as the middle element is moved over its range, to keep the focus exact. This kind of lens has come to be called a zoom lens.

By increasing the number of elements, the focus can be kept exact without the need of a correcting cam. Other improvements include increasing the range of focal lengths covered, increasing the effective lens aperture, increasing the angular field of view seen by the film, and improving the colour correction with radically new glass materials.

ART FOR PSYCHOTHERAPY

For a long time the change in focal lengths was carried out manually. More recently, the use of an electric motor drive has allowed a smoother change, with less distraction to the cameraman.

The general principles utilized in the film transport system have remained much the same over recent years, at least for the 35-mm film. The films are usually preloaded in light-tight reel cases (called magazines), with an exposed loop between the supply and take-up reels. This loop is quickly fitted into the camera mechanism when loading.

The intermittent is usually a claw-type mechanism, sometimes a "dual-fork" claw that pulls down four sprocket holes at a time. The fork protrudes and recedes to engage the sprocket holes. Some cameras are equipped with pin-registering mechanisms, which hold the film firmly in place in the exposure gate, with the pins engaging sprocket holes.

In the early days of sound films, the noise made by the intermittent and other moving parts in the camera was loud enough to interfere with the sound picked up by the microphone. Cameras were sheathed ("blimped") with outer, separate sound-absorbing materials. The sound insulation is now usually self-contained in the camera.

Before the introduction of sound, the film and intermittent were driven by a crank operated by the cameraman. With sound, considerably more uniformity in the speed of the film drive became necessary. For this and other reasons, the film drive in modern cameras is provided by an accurately controlled electric motor, which maintains the standardized sound speed of 24 frames per second.

ART FOR PSYCHOTHERAPY

The shutter keeps light from striking the film while it is moving from one frame to the next. A variable shutter opening can also be used to reduce exposure when it is necessary or desirable to do this without reducing the lens aperture. The shutter is in most cases rotary and is synchronized with the intermittent.

Viewfinding for motion pictures is especially critical: whereas still photographs can be cropped during enlargement or printing, the film image must be framed as it will appear on the screen. Older cameras employed a mechanical "rack-over" that enabled the camera operator to sight directly through the aperture with the film transport out of the way.

When an external viewfinder is used, the image seen through it is not exactly the same as that photographed. The viewfinder must be angled so that it and the taking lens both point at the centre of the subject. A system of cams in the focus mechanism of the camera keeps the viewfinder image free of parallax (viewpoint difference) by adjustment from infinity to the near-point of the lens with a separate cam for each focal length.

Most cameras used today are of the reflex type. A partially reflecting mirror (beam splitter) is positioned in the door of the camera body or built into the lens itself with a parallel viewing tube. The mirror diverts to the viewfinder some of the light rays coming through the lens. This method's major drawback is that it takes away part of the light that would otherwise be used for the exposure.

A much-admired viewing system that allows the full amount of light to reach the film is the rotating mirror shutter employed in the Arriflex camera. Light is reflected

ART FOR PSYCHOTHERAPY

into the viewfinder only when the shutter blade covers the film as it advances to the next frame. This arrangement, however, is not wholly free from objections. Chief among these is that the arrangement opens a return path for light from the viewer's eyepiece to reach the film. The eyepiece must fit snugly around the eye while the viewfinder is in use, and the finder must be closed completely while it is not in use.

In addition, since the camera shutter is closed only once per frame, the image will be subject to a distinct flicker, to which the cameraman must adjust himself. Some cameras incorporate a "video assist" or "video tap" wherein the viewfinder image is electronically fed to a video monitor or video recorder, thus allowing evaluation of the take by videotape replay.

Focusing has also been a perennial problem for the motion-picture camera. On the camera the position of the lens is precisely indicated on a calibrated scale. The actor's location on the set was formerly marked on the floor and the exact distance to the camera measured with a tape. The actor moved to previously marked places, and an assistant to the cameraman, called a focus puller, or follow-focus assistant, kept the lens in adjustment.

Various electrical devices have now been introduced for remote adjustment by the assistant. Where a through-the-lens finder is used, focusing can be done directly, using the viewfinder image. Also, experienced cameramen can estimate distances quite closely.

It is usual to generate some kind of signal in synchronism with the intermittent when an auxiliary, magnetic-tape sound recorder is used, so that the sound record can later be synchronized exactly with the picture. The sync-

ART FOR PSYCHOTHERAPY

generator provides a record of the speed of the camera motor; each frame of picture causes 2.5 cycles of a 60-hertz pulse to be recorded on the sync-track of the sound tape.

A newer system is based on the "time code" originally developed for videotape. A separate generator uses a digital audio signal to provide each frame of film with its own number.

For each take the time code generator is set to zero; when the camera and film are running, the generator starts to emit numbers that represent "real-time" in hours, minutes, seconds, and frames. In one system, a light-emitting diode next to the camera aperture records the information as ordinary numbers that can be read by the eye; in others, the binary numbers are contained in a control surface of magnetic particles on the base side of the film.

One hundred feet of 35-mm film would be rendered in time code as 00:01:07:08, or one minute, seven seconds, eight frames. Corresponding information is recorded on the "address" track of the audio tape.

The time code's last two digits, which represent frames, go up to either 24 or 30. Material intended for theatres is photographed at the international sound projection speed of 24 frames per second. Material filmed for American television is often shot at 30 frames per second (in countries with 50 hertz AC power, the rate is 25 frames per second).

The camera is often supplied with electric motors to perform miscellaneous functions, such as to provide smooth rotation (panning) of the camera or to change the magnification in a zoom lens (or change lenses in a turret). The camera is normally provided with footage indicators to indicate the amount of film left unexposed and with frame counters used when it is desired to superimpose a second exposure.

There can also be an "inching knob" to reposition the film to a given frame for multiple exposures. When the camera is used at a speed different from standard, a tachometer may be provided to indicate the actual speed.

The cameras that have so far been described are for the standard 35-mm film. Cameras for 65-mm film are generally quite similar, though heavier. The 16-mm professional camera may differ from the 35-mm in the form of its case, in its use of a spring-operated film drive, and in its method of film loading, as a result of its development from a former amateur camera.

On the other hand it may be a smaller version and have the same features as the 35-mm model by the same manufacturer.

Camera supports

The camera must be mounted on a substantial support to avoid extraneous movements while film is being exposed. In its simplest form this is a heavy tripod structure, with sturdy but smooth-moving adjustments and casters, so that the exact desired position can be quickly reached. Often a heavy dolly, holding both the camera and a seated cameraman, is used.

This can be pushed or driven around the set. When shots from elevated positions are to be used, both camera and cameraman are carried on the end of a crane, also on a dolly. In some cases the assemblage is smoothly driven

ART FOR PSYCHOTHERAPY

to follow the action being pictured, such as movement along a street. If the surface being traversed is not smooth, rails, resembling train tracks, must be laid on the floor or ground for the dolly.

The camera may be freed from the tripod or dolly and carried by the operator by means of a body brace and gyroscope stabilizer. One such support is the Steadicam, which eliminates the tell-tale motions of the hand-held camera.

Film

Film types are usually described by their gauge, or approximate width. The 65-mm format is used chiefly for special effects and for special systems such as IMAX and Showscan. It was formerly used for original photography in conjunction with 70-mm release prints; now 70-mm theatrical films are generally shot in 35-mm and blown up in printing. With some exceptions the 35-mm format is for theatrical use, 16-mm for institutional applications, and 8-mm for home movies.

There are some minor differences in the shape of the sprocket holes in 35-mm film between negative and positive film. The first 8-mm film was made by using 16-mm film, punched with twice as many sprocket holes of the same size and shape. One side, to the middle line, was exposed in one direction.

The supply and take-up reels were then interchanged in the camera, and the other side was exposed in the other direction. After processing, the film was split into two strips, which were spliced into one. An improved version of 8-mm stock, called Super-8 film, was designed with the idea of reducing the sprocket-hole size and employing the space thus made available for a larger picture area.

Originally, the film base was some form of celluloid or cellulose nitrate. This material is highly flammable, and extensive precautions were required in projection rooms to avoid film ignition because of the proximity of the projector arc lamp to the film. In 1923, when 16-mm amateur film was introduced, cellulose acetate (or safety film), much less flammable than the nitrate, was used. It was not considered desirable to adopt it for professional 35-mm film, largely because it was inferior in strength and dimensional stability.

By the late 1930s an improved cellulose acetate safety film was introduced, and by the early 1950s it had generally replaced the nitrate film. Since 1956 acetate has lost ground to polyester- or mylar-based film, which is thinner, less brittle, and more resistant to tearing.

Lighting

The art of cinematography is, above all, the art of lighting, and the British term for the chief of the camera crew, lighting cameraman, comes closer to the matter than the Hollywood director of photography. In motion-picture photography, decisions about exposure are governed by the overall style of film, and light levels are set to expose the particular film stock at the desired f-stop.

Light sources

The earliest effective motion-picture lighting source was natural daylight, which meant that films at first had to be photographed outdoors, on open-roof stages, or in glass-enclosed studios. After 1903, artificial light was introduced in the form of mercury vapour tubes that produced a rather flat lighting.

Ordinary tungsten (incandescent) lamps could not be used because the light rays they produced came predominantly from the red end of the spectrum, to which the orthochromatic film of the era was relatively insensitive. After about 1912, white flame carbon arc instruments, such as the Klieg light (made by Kliegl Brothers and used for stage shows) were adapted for motion pictures.

After the industry converted to sound in 1927, however, the sputtering created by carbon arcs caused them to be replaced by incandescent lighting. Fresnel-lens spotlights then became the standard. Fresnel lenses concentrate the light beam somewhat and prevent excessive light loss around the sides.

They can also, when suitably focused, give a relatively sharp beam. In the studio there are racks above and stands on the floor on which lamps can be mounted so that they direct the light where it is wanted.

The advent of Technicolor led to a partial reversion to the carbon arc because incandescent light affected the colours recorded on the film. Around 1950, however, economic pressures caused Technicolor film to be rebalanced for incandescent light.

The modern era in lighting began in the late 1960s when tungsten-halogen lamps with quartz envelopes came into wide use. The halogen compound is included inside the envelope, and its purpose is to combine with the tungsten evaporated from the hot filament.

ART FOR PSYCHOTHERAPY

This forms a compound that is electrically attracted back to the tungsten filament. It thus prevents the evaporated tungsten from condensing on the envelope and darkening it, an effect that reduces the light output of ordinary gas-filled tungsten lamps.

The return of the tungsten to the filament means that the incandescent lamp can be run with a long life at a higher filament temperature and, more important, remain at precisely the same colour temperature.

These lamps are now sometimes provided with a special multilayered filter to give a bluish light that approaches the colour of daylight. Halogen lamps give brilliant light from a compact unit and are particularly well-suited to location filming.

The principal light on a scene is called the key light. The position of the key light has often been conventionalized (*e.g.,* aimed at the actors at an angle 45 degrees off the camera-to-subject axis).

Another school of cinematographers prefers source lighting, in the tradition of Renaissance and Old Master paintings; that is, a window or lamp in the scene governs the angle and intensity of light. A fill light is used to provide detail in the shadow areas created by the key light.

The difference in lighting level between the key plus the fill light versus the fill light alone yields the lighting contrast ratio. The "latitude" of the film, or the spread between the greatest and least exposure that will produce an acceptable image, governs the lighting contrast ratio.

For many years, the latitude of colour films was so restricted that it was thought necessary to have numerically low lighting ratios, typically 2 to 1 (a very flat lighting) and never more than 3 to 1. The introduction of Eastman 5254 colour negative in 1968 and the even more sophisticated 5247 in 1974 opened a new era in which colour film was exposed with higher ratios approaching the previous subtleties of black-and-white.

Light measurement

Precise control of exposure throughout filming is necessary to maintain consistent tones from shot to shot and to give an overall tenor of lighting that suits the pictorial style. To determine light levels in the studio and on interior locations, an incident light meter is primarily used.

This type of meter is recognizable by a white plastic dome that collects light in a 180-degree pattern (the dome is an approximation of the shape of the human face). Because it measures the overall light (calibrated in foot-candles) falling on the scene, it may be used without the actors present.

Reflected light readings measure the average light coming toward the camera from the scene being photographed. This works well for average subjects but gives wrong exposures if the background contains either many bright areas, as in a beach scene, or very dark areas, as in front of a dark building. In such cases the photocell must be held not at the camera but very close to the subject of interest, to eliminate the effect of the background.

ART FOR PSYCHOTHERAPY

This is also the case when the scene contains a good deal of backlight. These shortcomings eventually led to the development of the spot meter.

Spot measurement readings measure the light coming toward the camera from selected spots in the subject being photographed. The meter for this purpose has an optical system that covers measurement of a spot of about one degree, making it extremely useful on exterior locations.

Light is also measurable in terms of colour temperature. Light rich in red rays has a low reading in kelvins. Ordinary household light bulbs produce light of about 2,800 kelvins, while daylight, which is rich in rays from the blue end of the spectrum, may have readings from 5,000 to more than 20,000 K.

The colour temperature meter uses a rotating filter to indicate a bias toward either red or blue; when red and blue rays are in balance, the needle does not move. Some meters also use red/blue and blue/green filters for fuller measurement.

The general practice has been to shoot the entire picture on stock balanced for artificial light at 3,200 K. Lights for filmmaking generally ranges between 3,200 K and 3,400 K. For daylight shooting, an orange filter is employed to counter the film's sensitivity to blue light.

Although colour-correcting filters are produced in a great many gradations, the No. 85 filter is generally used to shoot tungsten-balanced colour film outdoors. For mixed-light situations where daylight enters through windows but tungsten light is used for the interior, the practice has been to cover the windows with sheets of plastic similar in colour to the No. 85 filter.

ART FOR PSYCHOTHERAPY

This reduces the colour temperature of the natural light to that of the artificial light. When the windows are very large, blue filters are sometimes placed on the lights and the No. 85 orange filter is used on the lens, as if filming in exterior daylight.

Yet another approach is to supplement natural daylight with metal halide (daylight-balanced) lights. With the increase in location shooting, daylight-balanced high-speed films have been introduced to allow shooting in mixed-light situations without light loss due to filters.

Film processing and printing

In the early days of motion pictures, films were processed by winding on flat racks and then dipping in tanks of solution.

As films became longer, such methods proved to be too cumbersome. It was recognized that the processing system should have the following characteristics: it must run continuously; it must be light-tight and yet capable of being loaded in daylight; and it must be as compact as possible to provide a minimum air surface for the processing solutions. A general form evolved that is still in use.

For continuous operation the film must be passed continuously through the solutions and folded back over rollers that do not touch the emulsion surface. It must be handled very carefully, as the impregnation with solution weakens the support, and the sprocket holes should not be engaged. Drive should, therefore, be accomplished by a light friction force at the edges.

ART FOR PSYCHOTHERAPY

Splicing on a fresh film without affecting the motion of the part of the film being processed is handled by using a storage unit or reservoir. This reservoir has a variable capacity so that the output end can be giving out film while the input end is stationary as the new film is spliced.

Light tight gates prevent all but a short length of film being light-struck at the very beginning or end of the film (and leaders may be used).

The take-up-reel case is fastened in a light tight way to the storage unit so that after splicing, the film is unreeled into the storage and processing units until the other end is reached, ready for splicing to the next film after changing cases.

Many tank shapes have been tried. Long vertical tanks provide for several passes of the film through each tank. The spools are designed so as to hold the film at the edges by friction. There are a number of types of drive, but all function gently to avoid strain. Sometimes the spools have multi-stepped edges to accommodate various film widths.

The lower spools (or "diabolos") are more or less free but guided in a loose fashion so that they will not jam or tangle. The long vertical tanks give a minimum of air surface to the solution. The motion of the film through the liquid can be sufficient for proper contact of the film with the solutions, but sometimes submerged sprays with small jets of fine nitrogen bubbles are provided to increase the agitation.

The last receptacle in the processing sequence is a drying oven. There are several designs, some of which generally resemble the tank but without solution and are

ART FOR PSYCHOTHERAPY

provided with heating elements. This receptacle does not need to be light-tight.

The processing steps for the many different types of film are similar in principle, though there are variations in specific solutions and treatments. One variation is known as reversal processing. After partial development, the camera original is bleached and given a second exposure of uniform white light. This yields a positive rather than a negative image and thus saves the cost of an additional generation.

In laboratory parlance, the major functions are divided into "front end" and "release print" work and may be performed at different facilities. Front end work begins even before shooting with tests by the cinematographer on the same film stocks that will be used for the production.

These will be used as a guide when takes from the camera negative that come in from each day's shooting are printed. A colour video analyzer reads the red, blue, and green records of the tests over a range of six f-stops to establish "printer lights." As desired, the work print may be "one light" (given uniform exposure) or "timed" (exposure corrected for scene-to-scene variations).

The original negative is stored until postproduction is finished. Positive work print is furnished in 1,000-foot rolls for editing.

When all editing, including the insertion of optical effects and titles, is completed, the negative cutter matches the original camera film frame by frame at each editing point. The edited camera negative is combined with the synchronized sound track negative into a composite print called the answer print.

ART FOR PSYCHOTHERAPY

(The first answer print is rarely the same as the final release print.)

After all colour-correction and timing takes place, the information is recorded on perforated paper tape that serves to control both the exposure for each shot and the louvered filters that add red, green, and blue values.

For theatrical distribution, exhibition release prints are not normally struck from the original camera negative.

The original negative is used to make a master positive, sometimes known as the protection positive, from which a printing negative is then made to run off the release prints. Alternatively, a "dupe" negative can be made by copying the original camera negative through the reversal process. This yields a colour reversal intermediate (CRI) from which prints can be struck.

Printing takes a number of different forms. In contact printing, the master film (or negative) is pressed against the raw stock; this combination is exposed to light on the master film side. In optical printing, the master film is projected through a lens to expose the raw stock. In continuous printing, the master film and the raw stock both run continuously.

Continuous printing is usually contact printing but can be optical, through a projected slit. In intermittent, or step-by-step, printing, each frame of the master film is exposed as a whole to a corresponding frame space on the raw film.

It is possible to print from one size master film to another size raw stock, such as 35-mm to 16-mm, or vice versa. In such cases the printing must, of course, be optical, and in the examples cited must be intermittent if there is a sound track. This is because 35-mm sound film has a spacing between frames and 16-mm does not.

The sound track must be printed separately. The preferred method for making 16-mm versions of 35-mm films is to make a 16-mm negative by reduction from the 35-mm negative. Sometimes a 35-mm release print is reduced and printed by reversal, but this yields a coarser image.

When 16-mm film is "blown up," the 16-mm negative is immersed in a solution that conceals scratches and grain as it is being re-photographed; this process is called wet-gate printing.

Film prints to be used for projection are given a coat of wax over the sprocket-hole areas. This eases the film passage between the pressure plates at the projection aperture.

Sound-recording techniques

The art of sound recording for motion pictures has developed dramatically. Most of the improvements fall into three areas: fidelity of recording; separation and then resynchronization of sound to picture; and ability to manipulate sound during the postproduction stage.

Optical recording

Until the early 1950s the normal recording medium was film. Sound waves were converted into light and recorded onto 35-mm film stock. Today the principal use of optical recording is to make a master optical negative for final

ART FOR PSYCHOTHERAPY

exhibition prints after all editing and rerecording have been completed.

Magnetic recording

Magnetic recording offers better fidelity than optical sound, can be copied with less quality loss, and can be played back immediately without development. Magnetic tracks were first used by filmmakers in the late 1940s for recording music.

The physical principles are the same as those of the standard tape recorder: the microphone output is fed to a magnet past which a tape coated with iron oxide runs at a constant speed. The changes in magnetic flux are recorded onto the tape as an invisible magnetic "picture" of the sound.

At first the sound was recorded onto 35-mm film that had a magnetic coating. Today sprocketed 35-mm magnetic tape is used during the editing stages. For onset recording, however, the film industry converted gradually to the same un-perforated quarter-inch tape format widely used in broadcasting, the record industry, and even the home.

Documentary and independent filmmakers were the first to develop and use the portable, more compact apparatus. Improvements in magnetic recording have paralleled those in the recording industry and include the development of multiple-track recording and Dolby noise reduction.

ART FOR PSYCHOTHERAPY

Double-system recording

Although it is possible to reproduce sound, either optically or magnetically, in the same camera that is photographing a scene (a procedure known as single-system recording), there is greater flexibility if the sound track is recorded by a different person and on a separate unit.

The main professional use for single-system recording is in filming news, where there is little time to strive for optimal sound or image quality. Motion-picture sound recording customarily uses a double system in which the sound track remains physically separate from the image until the very last stages of postproduction.

Double-system shooting requires a means of re-matching corresponding sounds and images. The traditional solution is to mark the beginning of each take with a "clapper," or "clap-stick," a set of wooden jaws about a foot long, snapped together in the picture field. The instant of clacking then is registered on both picture and sound tracks.

Each new take number is identified visually by a number on the clapper board and aurally by voice. A newer version of the clapper is a digital slate that uses light-emitting diodes and an audio link to synchronize film and tape.

Precise synchronism must be maintained between camera and recorder so that sound can be kept perfectly matched to the visuals. (Lack of perfect synchronism is most conspicuous in close-up shots in which a speaker's lips do not match his voice.)

ART FOR PSYCHOTHERAPY

On some occasions several cameras shoot a scene simultaneously from different points of view while only one sound recording is made, or several sound records may be taken of a single shot. Thus, to maintain synchronism, all sound and picture versions of a particular scene must be recorded at the same speed; the camera and the recorder cannot fluctuate in speed.

One way to achieve this is to drive all cameras and recorders from a common power supply. Alternatively, synchronization may be achieved through the automatic, continual transmission from cameras to recorders of a sync-pulse signal sent by cable or wireless radio.

More convenient yet is crystal sync, whereby the speed of both cameras and recorders is controlled through the use of the oscillation of crystals installed in each piece of equipment. The most advanced system uses a time-code generator to emit numbers in "real-time" on both film and tape.

The sound recordist

The main task of the recordist during live recording is to get "clean" dialogue that eliminates background noise and seems to correspond to the space between speaker and camera. Most of the non-synchronous dialogue, sound effects, and music can be added and adjusted later.

During shooting the sound recordist adjusts the sound by setting levels, altering microphone placement, and mixing (combining signals if there is more than one microphone). Major technical and aesthetic reshaping is left for the postproduction phase when overhead is lower, the

ART FOR PSYCHOTHERAPY

facilities are more sophisticated, and alternative versions can be created.

It is also the job of the sound personnel to record wild sound (important sound effects and non-synchronous dialogue) and ambient sound (the inherent sound of the location). Ambient sound is added to the sound track during postproduction to maintain continuity between takes. Usually, wild sound and music are also adjusted and added then.

Microphones

Microphones of many different types have been used for sound recording. These may differ in sound quality, in directional characteristics, and in convenience of use.

Conditions that may dictate the choice of a particular microphone include the presence of minor echoes from objects in the set or reproduction of speech in a small room, as distinct from that in a large hall. Painstaking adjustments are made by careful attention to the choice of microphones, by the arrangement and sound absorbency of walls and furniture on the set, and by the exact positioning of the actors.

For recording a conversation indoors, the preferred microphone is sensitive in a particular direction in order to reduce extraneous noises from the side and rear. It is usually suspended from a pole-like "boom" just beyond camera range in front of and above the actors so that it can be pivoted toward each actor as he speaks. Microphones can also be mounted on a variety of other stands.

ART FOR PSYCHOTHERAPY

A second way to cut down background noise is to use a chest (or lavaliere) microphone hidden under the actor's clothing. For longer shots, radio microphones eliminate the wires connecting actors to recorders by using a miniature transistor radio to send sound to the mixer and recorder.

Newer techniques

Efforts to lessen the extraordinary labour and costs of artistic presentations and animation have taken two basic directions: simplification and computerization. Inexpensive cartoons made for television have often resorted to "limited animation", in which each drawing is repeated anywhere from two to five times.

The resultant movements are jerky, rather than smoothly gradated.

Often only part of the body is animated, and the background and the remaining parts of the figure do not change at all. Another shortcut is "cycling", whereby only a limited number of phases of body movement are drawn and then repeated to create more complicated movements such as walking or talking.

Although computers can be used to create the limited animation described above, they can also be used in virtually every step of sophisticated animation.

Computers have been used, for example, to automate the movement of the rostrum camera or to supply the in-between drawings for full animation.

If a three-dimensional figure is translated into computer terms (*i.e.,* digitized), the computer can move or rotate the object convincingly through space.

ART FOR PSYCHOTHERAPY

Hence, computer animation can demonstrate highly complex movements for medical or other scientific researchers.

Animators who work with computers usually distinguish between computer-assisted animation, which uses computers to facilitate some stages of the laborious production process, and computer-generated animation, which creates imagery through mathematical or computer language rather than through photography or drawing.

Finally, computers may be used to modify or enhance a drawing that has been initiated in the traditional manner.

ART FOR PSYCHOTHERAPY

SECTION SEVEN: ART THERAPY AND CONCEPTS

Memory

Memory (in biology and psychology), is the capacity of animals, including humans, to store information in the brain because of learning.

When an animal learns to associate two events, such as a stimulus followed by food, connections are made in the brain, which form a representation of these two events and their relationship to each other.

This is what constitutes memory. It is an essential process, which enables animals to suit their behaviour to the environment surrounding them, making use of the information learned by retrieving it at a later stage and applying it to a similar situation.

The study of memory in humans falls into two subject areas, biology, and psychology. Psychologists have distinguished between two main types of memory. Short-term, or primary, memory is used for the temporary holding of information that is of current concern.

It has been associated with attention and the contents of consciousness, although now it is described as a 'working memory', by contrast with the long-term, or permanent, memory that contains inactive stored information which can be retrieved and placed into short-term holding if required for current thought processes.

Psychologists also distinguish different parts of memory in the long-term store, those for visual images, personal life-events, knowledge of word meanings, and skilled physical behaviour, for example.

ART FOR PSYCHOTHERAPY

It is thought certain that some long-term change in the brain's physiology must act as a mechanism for memory, possibly through changes in the connections between nerve fibres.

The mechanism of short-term memory is thought possibly to be electrical in nature, involving the temporary activation of information being currently processed or 'thought about' by the brain.

The study of memory disorders may throw light on some of these questions. A quite different, social, phenomenon, which may be called 'collective memory', is connected with the formation of myth and stereotype.

Freud, Sigmund (1856-1939),

Freud is the Austrian psychologist and psychotherapist who pioneered psychoanalysis.

Following a medical training, in which he specialized in neurology, he turned to the study of hysteria. *Studies on Hysteria* (1895), written with Josef Breuer (1842-1925), established the framework of psychoanalytic theories about neurosis: that the symptoms result from (and have a symbolic relation to) an emotional trauma, which the patient has 'forgotten'.

The memory, however, acts in the unconscious to disrupt the patient's thoughts and feelings. To gain access to unconscious material the therapist uses hypnosis, dream analysis, or free association.

Freud published *The Interpretation of Dreams* in 1900. He noticed the importance of sexual themes in this material and argued that sexual 'memories' were really childhood fantasies, which represented infantile sexual desires.

ART FOR PSYCHOTHERAPY

This thesis of childhood sexuality, laid out in *Three Essays on the Theory of Sexuality* (1905), became central to Freud's thinking about normal and abnormal psychosexual development. His theories came from clinical observations about which he wrote case histories; some of which, however, were subsequently shown to have been self-censored.

There are many sources of distortion in such data and Freud's preoccupation with sexual explanations sometimes led him to ignore more obvious possibilities. Freudian theory enjoyed great popularity in the 1940s and 1950s, when academics such as the social learning theorists at Yale University in the USA attempted to test his ideas by experiment, but they were rarely confirmed.

Freud and his followers developed such key concepts as displacement, identification, projection, regression, repression, and sublimation (see defence mechanisms). His ideas remain controversial and influential not only in the treatment of psychiatric illness, but in the arts and social sciences. Later psychoanalytic theorists who follow Freud are known as neo-Freudian

Psychotropic drug

Psychotropic drug is any member of a large group of substances used in the treatment of psychological disorders.

They are classified according to the conditions for which they are given, and each class is further divided according to the chemical nature of the drug.

Major tranquillizers, such as phenothiazines, may be given for severe psychotic disorders such as

ART FOR PSYCHOTHERAPY

schizophrenia and mania; they are sometimes called neuroleptics.

Minor tranquillizers, of which the most popular are the benzodiazepines, are effective in anxiety states.

Antidepressants are used to relieve severe depression and are classified according to their chemical structure (for example, tricyclic antidepressants) or their biochemical actions (for example, monoamine oxidase inhibitors).

The mode of action of psychotropic drugs may give information about the nature of the disease in which they are effective.

For example, antidepressants are known to increase the concentration of certain chemical transmitters at the ends of nerves in the brain.

This fact has led to the hypothesis that depression is caused by an abnormally low concentration of such transmitters.

Many such hypotheses depend on an over-simplified view of an exceedingly complicated subject, and have not stood the test of time.

Nevertheless, the discovery and marketing of psychotropic drugs has been an important branch of pharmaceutics over the last twenty years.

Psychotropic drugs are sometimes used in conditions other than mental disease: for example, the neuroleptic chlorpromazine is effective in stopping severe vomiting, and the benzodiazepines may be used to treat some forms of epilepsy.

ART FOR PSYCHOTHERAPY

Tranquillizer

Tranquillizer is a drug having a calming effect. Tranquillizers calm and relax patients with less sedative and hypnotic actions than other drugs. 'Major' tranquillizers (for example, chlorpromazine) are used to treat severe psychotic disorders such as mania and schizophrenia.

'Minor' tranquillizers are effective against anxiety and insomnia, and have largely superseded barbiturates; an example is diazepam, the best-known formulation of which is Valium.

Minor tranquillizers are less liable to produce dependence than barbiturates, and overdoses are far less dangerous

Remembering

This is the encoding, storage, and retrieval in the human mind of past experiences.

The fact that experiences influence subsequent behaviour is evidence of an obvious but remarkable activity called remembering. Memory is both a result of and an influence on perception, attention, and learning.

The basic pattern of remembering consists of attention to an event followed by the representation of that event in the brain. Repeated attention, or practice, results in a cumulative effect on memory and enables activities such as a skilful performance on a musical instrument, the recitation of a poem, and reading and understanding words on a page.

Learning could not occur without the function of memory. So-called intelligent behaviour demands memory,

ART FOR PSYCHOTHERAPY

remembering being prerequisite to reasoning. The ability to solve any problem or even to recognize that a problem exists depends on memory. Routine action, such as the decision to cross a street, is based on remembering numerous earlier experiences.

The act of remembering an experience and bringing it to consciousness at a later time requires an association, which is formed from the experience, and a "retrieval cue," which elicits the memory of the experience.

Practice (or review) tends to build and maintain memory for a task or for any learned material. During a period without practice, what has been learned tends to be forgotten. Although the adaptive value of forgetting may not be obvious, dramatic instances of sudden forgetting (as in amnesia) can be seen to be adaptive.

In this sense, the ability to forget can be interpreted as having been naturally selected in animals. Indeed, when one's memory of an emotionally painful experience leads to severe anxiety, forgetting may produce relief.

Nevertheless, an evolutionary interpretation might make it difficult to understand how the commonly gradual process of forgetting was selected for.

In speculating about the evolution of memory, it is helpful to consider what would happen if memories failed to fade. Forgetting clearly aids orientation in time; since old memories weaken and new ones tend to be vivid, clues are provided for inferring duration.

Without forgetting, adaptive ability would suffer; for example, learned behaviour that might have been correct a decade ago may no longer be appropriate or safe.

ART FOR PSYCHOTHERAPY

Indeed, cases are recorded of people who (by ordinary standards) forget so little that their everyday activities are full of confusion. Thus, forgetting seems to serve the survival not only of the individual but of the entire human species.

Additional speculation posits a memory-storage system of limited capacity that provides adaptive flexibility specifically through forgetting. According to this view, continual adjustments are made between learning or memory storage (input) and forgetting (output).

There is evidence in fact that the rate at which individuals forget is directly related to how much they have learned. Such data offer gross support for models of memory that assume an input-output balance.

Whatever its origins, forgetting has attracted considerable investigative attention. Much of this research has been aimed at discovering those factors that change the rate of forgetting.

Efforts are made to study how information may be stored or encoded in the human brain. Remembered experiences may be said to consist of encoded collections of interacting information, and interaction seems to be a prime factor in forgetting.

Memory researchers have generally supposed that anything that influences the behaviour of an organism endowed with a central nervous system leaves—somewhere in that system—a "trace" or group of traces. So long as these traces endure, they can, in theory, be re-stimulated, causing the event or experience that established them to be remembered.

ART FOR PSYCHOTHERAPY

Time-dependent aspects of memory

Research by the American psychologist and philosopher William James (1842–1910) led him to distinguish two types of memory: primary, for handling immediate concerns, and secondary, for managing a storehouse of information accumulated over time.

Memory researchers have since used the term *short-term memory* to refer to the primary or short-lived memory functions identified by James. *Long-term memory* refers to the relatively permanent information that is stored in and retrieved from the brain.

Working memory

Some aspects of memory can be likened to a system for storing and efficiently retrieving information. One system in particular—identified as "working memory" by the British psychologist Alan Baddeley—is essential for problem solving or the execution of complex cognitive tasks. It is characterized by two components: short-term memory and "executive attention."

Short-term memory comprises the extremely limited number of items that humans are capable of keeping in mind at one time, whereas executive attention is a function that regulates the quantity and type of information that is either accepted into or blocked from short-term memory.

Baddeley likened working memory to a scratch pad in which essential pieces of information are inscribed and later discarded (or, as is more likely the case, replaced by more pertinent information).

ART FOR PSYCHOTHERAPY

Executive attention

In its role of managing information in short-term memory, executive attention is highly effective in blocking potentially distracting information from the focus of attention.

This is one way in which the brain is able to keep information active and in focus. Yet there are limits to the amount of information ("capacity") that executive attention is capable of handling at any given time, and this capacity will differ from person to person.

As a result, all people differ in their ability to bring attention to bear on the control of thought. Known as "working memory capacity," this ability is measured most often through a test that requires people to commit a short list of items to memory while performing some other task.

Thus, one form of the test might involve reading a series of sentences and then attempting to recall the letters at the end of each sentence. The capacity of working memory is measured by the number of items that a person recalls, so that if a person recalls five letters, the working memory capacity in this case is five.

In most cases, number of letters recalled will depend on each person's ability to avoid the distraction of reading the sentences. Such tests of working-memory capacity can be used to predict an individual's ability to perform tasks involved in reasoning. In fact, working memory capacity is strongly related to general intelligence.

In terms of brain activity, executive attention seems to involve the frontal lobes. Thus, damage to the frontal lobes, which is associated with a condition called dysexecutive syndrome, can affect the role of executive attention in the control of thought, behaviour, and emotion.

Evidenced by a notable reduction in the patient's abilities to set goals, make plans, and initiate actions, dysexecutive syndrome is often accompanied by diminished social inhibitions and thereby leads to behaviour that is considered rude or inappropriate. Excessive use of alcohol and other drugs can lead to similar behavioural problems.

Patterns of acquisition in working memory

In the course of a typical day, humans receive a continuous stream of information from the world around them as well as from their own thought processes and physical experiences. They manage this constant stimulation through a combination of conscious and unconscious effort.

The majority of the information is processed (or ignored) unconsciously, because the brain is incapable of consciously attending to and filtering every bit of stimulation it receives. Other forms of information that are processed through unconscious effort, such as a loud sound or a sudden bright light, tend to capture attention in various ways.

While events that elicit such attention are more likely to be remembered, especially if they need to be retrieved for possible use in the future, the more significant processes of conscious attention are volitional, occurring in everyday actions such as driving, reading a book, playing chess, watching a basketball game, and following a recipe in a cookbook.

The level of attention given to an experience, and the way a person thinks about it, will influence how well the memory for the event is acquired and how well it will be recalled. Researchers also have determined that the techniques employed by the brain in acquiring information differ depending on whether the information is intended for short-term or long-term use.

Most people are capable of storing a maximum of about seven separate units of information in short-term memory—e.g., the seven random letters F, L, I, X, T, Z, R. Thus, one may consult a directory for a 10-digit telephone number but forget some of the digits before one has finished dialling.

However, if the units of information are grouped or "chunked" into meaningful patterns, it is possible to recall many more of them, as shown by the series of 24 letters F, R, A, N, C, E, G, E, R, M, A, N, Y, P, O, L, A, N, D, S, P, A, I, N. According to the American psychologist George A. Miller, such chunking of information is essential for short-term memory and plays an important role in learning.

Short-term memory is restricted in both capacity and duration: a limited amount of information will remain active for a few seconds at best unless renewed attention to the information successfully reactivates it in working memory. Before such "renewal" occurs, most information arrives in working memory through sensory inputs, the two most prevalent being aural and visual.

Baddeley posited that working memory is supported by two systems: the phonological loop, which processes

ART FOR PSYCHOTHERAPY

aural information, and the visuo-spatial sketch pad, which processes visual and spatial information. When information is acquired aurally, the brain encodes the information according to the way it sounds.

A person who hears a spoken telephone number and retains the information long enough to complete dialling is employing the phonological loop, a function of working memory involving, in effect, an inner voice and inner ear each person utilizes to rehearse and recall information. Children who are slow to learn this type of encoding are also generally delayed in learning to read.

Visual and spatial encoding are an integral part of daily problem solving. A person solves a jigsaw puzzle by constructing an image of a missing piece and then seeking the piece that matches the constructed image.

It would not make sense for this construct to be held in long-term memory, but its function as a short-term memory is essential to reaching a solution. Such short-term encoding of visual-spatial information is important in any number of tasks, such as packing suitcases in the trunk of a car or searching for a missing shoe in the bottom of a closet.

Long-term memory

Memories that endure outside of immediate consciousness are known as *long-term memories*.

They may be about something that happened many years ago, such as who attended one's fifth birthday party, or they may concern relatively recent experiences, such as the courses that were served at a luncheon earlier in the day.

ART FOR PSYCHOTHERAPY

Accumulated evidence suggests that a long-term memory is a collection of information augmented by retrieval attributes that allow a person to distinguish one particular memory from all of the other memories stored in the brain. The items stored in long-term memory represent facts as well as impressions of people, objects, and actions.

They can be classified as either "declarative" or "non-declarative", depending on whether their content is such that it can be expressed by a declarative sentence.

Thus, declarative memories, like declarative sentences, contain information about facts and events. Non-declarative memory, also known as procedural memory, is the repository of information about basic skills, motor (muscular) movement, verbal qualities, visual images, and emotions.

A crosscutting distinction is made between memories that are tied to a particular place and time, known as "episodic" memories, and those that lack such an association, known as "semantic" memories.

The latter category includes definitions and many kinds of factual knowledge, such as knowledge of the name of the current pope, which one might not recall having learned at any particular time or place.

Patterns of acquisition in long-term memory

There are roughly three phases in the life of a long-term memory. It must be acquired or learned; it must be stored or retained over time; and, if it is to be of any value, it must be successfully retrieved.

ART FOR PSYCHOTHERAPY

These three phases are known as acquisition, storage, and retrieval. Relatively little is known about the factors influencing the storage of memory over time, but a good deal is understood about the mechanisms by which memories are acquired and successfully retrieved.

Intervals

Memory researchers have identified specific techniques for improving one's ability to remember information over a long period of time. One of the most powerful means involves scheduling regular practice sessions over a relatively long period.

Consider, for example, two groups of people learning vocabulary words in a foreign language. One group studies for five hours on one day, and the other group studies for one hour per day for five days in a row.

Although both groups practice for a total of five hours, they will differ in their ability to recall what they have learned. If the two groups are tested on the day after the first group studied for five hours, the first group will perform better than the second; if, on the other hand, the test occurs one week after the two groups completed their study, the second group will perform better and remember more of the words in the future.

Such cases suggest that, while there may be some short-term benefit to "cramming" for a test, the most effective means of committing facts to long-term memory depends upon routine and repetitive study.

ART FOR PSYCHOTHERAPY

Rehearsal

Although the ability to commit information to memory is greatly enhanced through repetition or rehearsal, not all rehearsal techniques are effective in facilitating later recall. Simply saying something to oneself over and over again, a technique called "rote rehearsal," helps to retain the information in short-term memory but does little to build a long-term memory of the event.

Another form of rehearsal involves motor coordination, whereby movements or series of movements are "memorized" for greater efficiency or skill of execution in the future.

A skilled touch typist who frequently inputs a short string of letters might thereby encode the movements involved in typing the full string, rather than relying on the separate movements he has already encoded for each letter.

In this sense, rehearsal occurs through repeated attention to each of several movements in a series.

This form of rehearsal enables the performance of countless activities, such as riding a bicycle, dancing a *particular step, or executing a competitive dive.*

Mnemonic systems

More-effective types of rehearsal consist of reflection—thinking about the material one is trying to learn and discovering ways in which it is related to something one already knows.

One traditional technique for committing a list of items to memory involves imagining that one is travelling a

ART FOR PSYCHOTHERAPY

familiar route in one's town while stopping to place an image of each item at specific landmarks on the route.

This technique, called the method of loci, was used by Greek and Roman orators such as Cicero and Simonides as a means of organizing and remembering points in their speeches.

The method of loci is based on the principle that encoding new information—such as items from the list to be memorized—to previously stored data—landmarks along a familiar route in one's town—can be an effective means of improving memory function.

When encoding techniques are formally applied, they are called mnemonic systems or devices. (The popular rhyme that begins "Thirty days hath September" is an example.) Verbal learning can be enhanced by an appropriate mnemonic system.

Thus, paired associates (e.g., DOG-CHAIR) will be learned more rapidly if they are included in a simple sentence (e.g., the dog jumped over the chair). Imagery that can associate different words to be learned (even in a bizarre fashion) has been found beneficial.

Indeed, some investigators hold that pure rote learning (in which no use is made of established memories except to directly perceive the stimuli) is rare or nonexistent. They suggest that all learning elaborates on memories already available.

Factors that influence the rate of learning should be distinguished from those that affect the rate of forgetting. For example, nonsense syllables are learned more slowly than are an equal number of common words; if both are studied for the same length of time, the better-learned common words will be forgotten more slowly.

However, this does not mean that the rate of forgetting intrinsically differs for the two tasks. Degree of learning must be held constant before it may be judged whether there are differences in rate of forgetting; rates of forgetting can be compared only if tasks are learned to an equivalent degree.

Indeed, when degree of learning is experimentally controlled, different kinds of information are forgotten at about the same rate. Nonsense syllables are not forgotten more rapidly than are ordinary words.

In general, factors that seem to produce wide differences in rate of learning show little (if any) effect on rate of forgetting, though some studies of mnemonic systems have demonstrated that pictorial (visual) mnemonics are associated with longer-held memories.

Physiological aspects of long-term memory

Investigators concerned with the physiological bases of memory seek a kind of neuro-chemical code with enough physical stability to produce a structural change or memory trace (engram) in the nervous system; mechanisms for decoding and retrieval also are sought.

Efforts at the strict behavioural level similarly are directed toward describing encoding, decoding, and retrieval mechanisms as well as the content of the stored information.

One way to characterize a memory (or memory trace) is to identify the information it encodes. A learner may encode far more information than is apparent in the task as presented.

For example, if a subject is shown three words for a few seconds and, after 30 seconds of diversion or distraction, is asked to repeat the process of learning-delay-recall with three new word groups, poorer and poorer recall will be observed on successive trials in cases where all of the word groups share some common element (e.g., all are animal names).

Such findings may be explained by assuming that the learner encodes this animal category as part of his memory for each word. Initially, the common category might be expected to aid recall by sharply delimiting the number of probable words.

Successive triads, however, tend to be encoded in increasingly similar ways, blurring their unique characteristics for the subject. An additional step provides critical supporting evidence for such an interpretation. If a final triad of vegetable names is unexpectedly presented, recall recovers dramatically.

The person being tested will tend to reproduce the vegetable names much better than he does those of the last animal triad, and recall will be roughly as efficient as it was for the first three animal names. This shift in word category seems to provide escape from earlier confusion or blurring, and it may be inferred that a common conceptual characteristic was encoded for each animal name.

Any characteristic or attribute of a word may be investigated in this way to determine whether it is incorporated in memory. When recall does not recover, it

ART FOR PSYCHOTHERAPY

may be inferred that the manipulated characteristic has little or no representation in memory.

For example, grammatical class typically does not appear to be encoded; decrement in recall produced after a series of triads consisting of verbs tends to continue when a shift is made to adjectives.

Such an experiment does not indicate what common encoding characteristic might be responsible for the decrement, suggesting only that it is not grammatical class.

Encoding mechanisms also may be inferred from tests of recognition. In one kind of experiment, for example, subjects study a long list of words, being informed of a multiple-choice memory test to follow.

Each word is made part of a test question that includes other carefully chosen new words, or "distractors." Distractors are selected to represent the different types of encoding the investigator suspects may have occurred in learning.

If the word selected for study is chosen by the subject, little can be inferred about the nature of the encoding. Any errors, however, can be most suggestive.

Thus, if the word to be studied was TABLE, the multiple-choice list of words might be TABLE, CHAIR, ABLE, FURNITURE, PENCIL, with TABLE being the only correct answer.

If CHAIR is incorrectly selected, it may be suspected that this associatively related word occurred to the subject implicitly during learning and became so well encoded that the subject later could not determine whether it or TABLE had been presented for memorization.

ART FOR PSYCHOTHERAPY

If the wrong choice is ABLE, acoustical resemblance to TABLE may have contributed to the confusion. If FURNITURE is erroneously chosen, it is likely that the conceptual category was prominent in the encoding.

Finally, because it is not related in any obvious way to TABLE, the word PENCIL may be intended as a control, unlikely to be a part of the memory for TABLE. If this is the case, subjects will be more likely to select distractor words such as PENCIL (or any others that have been encoded along with TABLE).

It is important to note the limitations on what may be inferred from experiments of this kind. Although a subject may have encoded in ways suggested by particular distractors, he still may be able to choose the correct word. Or, even if he chooses one of the distractor words, he still may have encoded in ways not represented by that word.

Retrieval

The common experience of having a name or word on the tip of the tongue seems related to specific perceptual (e.g., visual or auditory) attributes. In particular, people who report a "tip-of-the-tongue" experience usually are able to identify the word's first letter and the number of syllables with an accuracy that far exceeds mere guessing.

There is evidence that memories may encode information about when they were established and about how often they have been experienced.

ART FOR PSYCHOTHERAPY

Some seem to embrace spatial information; e.g., one remembers a particular news item to be on the lower right-hand side of the front page of a newspaper.

Research indicates that the rate of forgetting varies for different attributes. For example, memories in which auditory attributes seem dominant tend to be more rapidly forgotten than those with minimal acoustic characteristics.

The Canadian psychologist Endel Tulving has demonstrated that, while information may be retained over a long period of time, there is no guarantee it will be retrieved precisely when it is needed. Successful retrieval is much more likely if a person is tested in a physical setting (context) that is naturally associated with the event or fact.

In cases where the context during the recall test differs from the setting in which the learning occurred, retrieval will be less likely. This is why the name of a colleague from school or work may be difficult to recall if one happens to encounter him at a shopping mall. In such cases, the new setting interferes with one's ability to retrieve the person's name from long-term memory.

Memory can be aided by any number of cues, however. It would be far easier to recall the colleague's name if one was asked to choose it from a list. In general, "recognition memory" (involved in choosing the correct answer from a list) is more reliable than recall memory (retrieving information without any clue or hint that could assist in the retrieval). For this reason, most students prefer multiple-choice tests to essay tests.

If a designated (target) memory consists of a collection of attributes, its recall or retrieval should be enhanced by

ART FOR PSYCHOTHERAPY

any cue that represents or suggests one of the attributes. A person who fails to recall the word *horse*, for example, may suddenly remember it when he is told that there was an animal name on the list of words he studied.

Alternatively, he may remember it when presented with an associated term such as *barn* or *zebra*. While recall can be enhanced somewhat by cues, failures are common even with cues that are highly relevant. In sum, if words were not encoded or stored in the brain with accompanying attributes at the time of learning, cuing of any kind would be ineffective.

Retrieval is also influenced by the way in which information is organized in memory. Although it is possible to name all of the Canadian provinces and territories by randomly recalling the names that come to mind, a far more reliable means would be to recall the information systematically, say by geographic region or by alphabetical order.

The passage of time is another phenomenon that influences the successful recall of stored information. If a person is asked to name the opponents his favourite football team played last season and the score of each game, his responses usually will be most accurate for the games played at the beginning and the end of the season.

Similarly, a person asked to describe each day of an extended journey will best retrieve his memory of events that occurred during the beginning and the end of the trip. In a similar test, a person who is asked to recall the words on a list he has just viewed will recall the initial words in the list best ("primacy effects") and those at the end next best ("recency effects"), while items from the middle are least likely to be recalled.

ART FOR PSYCHOTHERAPY

This outcome will be consistent as long as recall begins immediately following presentation of the last word. If, however, a short interval follows that prevents the subject from rehearsing the contents of the list, the recency effect may disappear completely, causing words at the end of the list to be recalled no better than those appearing in the middle.

Thus, while primacy effects remain essentially undisturbed, a delay as short as 15 seconds can abolish the recency phenomenon. Although some researchers have suggested that recency effects depend on a separate short-term memory system while primacy effects are mediated by a long-term system, it is possible that a single memory function influences these outcomes.

Relearning

The number of successive trials a subject takes to reach a specified level of proficiency may be compared with the number of trials he later needs to attain the same level. This yields a measure of retention by what is called the relearning method.

The fewer trials needed to reach the original level of mastery, the better the subject seems to remember.

The relearning measure sometimes is expressed as a so-called savings score. If 10 trials initially were required, and 5 relearning trials later produce the same level of proficiency, then 5 trials have been saved; the savings score is 50 percent (that is, 50 percent of the original 10 trials). The more forgetting, the lower the savings score.

Although it may seem paradoxical, relearning methods can yield both sensitive and insensitive measures of forgetting. Tasks have been devised that produce wide differences in recall but for which no differences in relearning are observed.

Some theorists attribute this to a form of heavy interference among learned data that has only momentary influence on retention.

Six months or a year after initial learning, some tests may give zero recall scores but can show savings in relearning.

This suggests a cumulative effect, whereby previously acquired knowledge enhances future learning.

Autobiographical memory

As an aspect of episodic memory, autobiographical memories are unique to each individual. The study of autobiographical memory poses problems, because it is difficult to prove whether the events took place as reported.

Using diary methods, researchers have found that people recall actions more accurately than thoughts—except in the case of emotionally charged thoughts, which are particularly well-remembered.

Although very few errors are made by those undergoing tests of autobiographical memory, any errors typically involve mixing the details of separate events into one episode.

Another method for testing autobiographical memory involves asking subjects to associate particular autobiographical memories with various cue words, such as *window* or *rain*.

ART FOR PSYCHOTHERAPY

Eyewitness memory

Conflicting accounts by eyewitnesses demonstrate that memory is not a perfect recording of events from the past; indeed, it is actually a reconstruction of past events.

A particularly striking demonstration of the inaccuracy of eyewitness testimony comes from dozens of cases in which those convicted of serious crimes were freed from prison because DNA evidence proved they were not guilty. In most of these cases, the individuals had been convicted on the basis of eyewitness testimony.

Many phenomena can degrade the accuracy of memories. For example, the memory of an eyewitness to a crime may be distorted if he reads news accounts of the crime that contain photographs of a person suspected of committing it.

Later, the eyewitness may erroneously believe that the suspect in the news account is the person whom he saw commit the crime.

In this case, memory of the crime and memory of the photograph blend to create a vivid—albeit incorrect— memory of an event that never occurred.

Such inaccuracies are not uncommon. The American psychologist Elizabeth Loftus showed that even the manner in which people are questioned about an event can alter their memory of it.

Other studies have shown that psychotherapists may inadvertently implant false memories in the minds of their clients. Such outcomes illustrate the degree to which imagination can have powerful effects on memory.

ART FOR PSYCHOTHERAPY

False memories also can be created in laboratory experiments. Subjects who are asked to study a list of words that are related to a particular non-presented word will claim to remember seeing the non-presented word.

For example, after studying the words *bed, rest, wake, tired, awake, dream, doze, blanket, snooze, drowsy, snore*, and *nap,* a large number of subjects will claim to recall seeing the word *sleep*, even though it was not on the list.

Although false memories created in laboratory settings differ from false memories of real-world events, they cast light on the processes involved in the creation and maintenance of memory errors, as demonstrated in research by the American psychologists Henry Roediger and Kathleen McDermott.

Forgetting

When a memory of a past experience is not activated for days or months, forgetting tends to occur. Yet it is erroneous to think that memories simply fade over time—the steps involved are far more complex.

In seeking to understand forgetting in the context of memory, such auxiliary phenomena as differences in the rates of forgetting for different kinds of information also must be taken into account.

It has been suggested that, as time passes, the physiological bases of memory tend to change. With disuse, according to this view, the neural engram (the memory trace in the brain) gradually decays or loses its clarity.

ART FOR PSYCHOTHERAPY

While such a theory seems reasonable, it would, if left at this point, do little more than restate behavioural evidence of forgetting at the nervous-system level.

Decay or deterioration does not seem attributable merely to the passage of time; some underlying physical process needs to be demonstrated.

Until a neuro-chemical basis for memory can be more explicitly described, any decay theory of forgetting must await detailed development.

Interference

A prominent theory of forgetting at the behavioural level is anchored in the phenomenon of interference, or inhibition, which can be either retroactive or proactive.

In retroactive inhibition, new learning interferes with the retention of old memories; in proactive inhibition, old memories interfere with the retention of new learning. Both phenomena have great implications for all kinds of human learning.

In a typical study of interference, subjects are asked to learn two successive verbal lists. The following day some are asked to recall the first list and others to recall the second. A third (control) group learns only one list and is asked to recall it a day later.

People who learn two lists nearly always recall fewer words than those in the control group.

Theorists attribute the loss produced by these procedures to interference between list-learning tasks. When lists are constructed to exhibit varying differences, the degree of interference seems to be related to the amount of similarity.

Thus, loss in recall will be reduced when two successive lists have no identical terms. Maximum loss generally will occur when there appears to be heavy (but not complete) overlap in the memory attributes for the two lists.

One may recall parts of the first list in trying to remember the second and vice versa. Discrimination tends to deteriorate as the number of lists increases, retroactive and proactive inhibition increasing correspondingly, and suggesting interference at the time of recall.

In retroactive inhibition, however, not all of the loss need be attributed to competition at the moment of recall. Some of the first list may be lost to memory in learning the second; this is called unlearning.

If one is asked to recall from both lists combined, first-list items are less likely to be remembered than if the second list had not been learned.

Learning the second list seems to act backward in time (retroactively) to destroy some memory of the first. Much effort has been devoted to studying the conditions that affect unlearning, which has become a major topic in interference theory.

Retroactive and proactive effects can be quite gross quantitatively. If one learns a list one day and tries to recall it the next, learns a second list and attempts recall for it the following day, learns a third, and so on, recall for each successive list tends to decline.

Roughly, 80 percent recall may be anticipated for the first list; this declines steeply to about 20 percent for the 10th

ART FOR PSYCHOTHERAPY

list. Learning the earlier lists seems to act forward in time (proactively) to inhibit retention of later lists.

These proactive phenomena indicate that the more one learns, the more rapidly one forgets. Similar effects can be demonstrated for retroactive inhibition within just one laboratory session.

Such powerful effects have led some researchers to speculate that all forgetting is produced by interference. Any given memory is said to be subject to interference from others established earlier or subsequently. Interference, theoretically, may occur when memories conflict through any attributes.

With a limited group of attributes and an enormous number of memories, it might seem that ordinary attempts at recall would be chaotic. Yet even if all of the memories shared some information, other attributes not held in common could still serve to distinguish them.

For example, every memory theoretically is encoded at a different time, and temporal attributes might serve to discriminate otherwise conflicting memories.

Indeed, when two apparently conflicting lists are learned several days apart, proactive inhibition is markedly reduced. Assuming that memories are multiply encoded, interference theory need not predict utter confusion in remembering.

Sources of interference are quite pervasive and should not be considered narrowly. For example, all memories seem to be established in specific surroundings or contexts, and subsequent efforts to remember tend to be less effective when the circumstances differ from the original.

ART FOR PSYCHOTHERAPY

Alcoholics, when sober, tend to have trouble finding bottles they have hidden while intoxicated; when they drink again, the task is much easier. Some contexts also may be associated with other memories that interfere with whatever it is that one is trying to remember.

Each new memory tends to amalgamate information already in long-term storage. Encoding mechanisms invariably adapt or associate fresh data to information already present, to such an extent that what is encoded may not be a direct representation of incoming stimuli.

This is particularly apparent when the input is relatively meaningless; the newly encoded memory comes to resemble those previously established (i.e., it accrues meaning). For example, a nonsense word such as LAJOR might be encoded as MAJOR.

Challenges to interference theory

Although interference theory has attracted wide support as an account of forgetting, it must be placed in perspective.

Interpretations that emphasize distinctions between short- and long-term memory and that posit control processes for handling information are potentially more comprehensive than interference theory, and the behavioural evidence for interference eventually may be explained within such systems.

In addition, a number of predictions derived from interference theory have not been well supported by experiment. The focus of difficulty lies in the hypothesis that interference from established memories is a major source of proactive inhibition.

ART FOR PSYCHOTHERAPY

The laboratory subject is asked to learn tasks with attributes that have varying degrees of conflict with memories established in daily life.

Theoretically, the more conflict, the greater the proactive interference to produce forgetting. Yet a number of experiments have failed to provide much support for this prediction.

Interference theory also fails to account for some pathological forms of forgetting. Repression as observed in psychiatric practice, for example, represents almost complete, highly selective forgetting, far beyond that anticipated by interference theorists.

Attempts to study repression through laboratory procedures have failed to yield systematic data that could be used to test theoretical conclusions.

Amnesia

If humans forgot everything, the consequences would be devastating to their daily lives. It would be impossible to do one's job—much less find one's way to work.

Individuals who suffer damage to certain brain regions, particularly the hippocampus, experience this kind of significant memory loss, amnesia, which is marked by an inability to create new long-term memories.

In addition, some amnesiacs lose their ability to recall events that occurred before the brain injury, a condition known as retrograde amnesia.

Some amnesiacs do not experience deficits in short-term memory, and in many cases, their memory deficits appear

ART FOR PSYCHOTHERAPY

to be limited to the acquisition and recollection of new associations.

If an amnesic is introduced to a new acquaintance that leaves the room and returns a few minutes later, the amnesic will not remember having met that person. However, amnesiacs are able to remember some types of new information, though they may be unaware that they are remembering.

This was proved in the early 20th century by the French physician Édouard Claparède, who used a pin to prick an amnesic woman each time he shook her hand. Later the patient would not shake hands with Claparède, even though she could not readily explain why.

In her case, procedural memory effectively helped her avoid the physical pain that accompanied every act of shaking hands with the physician. Such studies demonstrate that procedural memory can function independently of conscious awareness.

Another form of forgetting is associated with the earliest stages of human development: nearly all people lack the ability to retain memories of experiences they had before they were three years old. Known as infantile amnesia, this universal phenomenon implies that the brain systems required to encode and retrieve specific events are not adequately developed to support long-term memory before age three.

Another theory points to developmental changes in the means by which memories are formed and retrieved after early childhood, suggesting that the more-developed brain lacks the ability to access such early memories.

ART FOR PSYCHOTHERAPY

Sigmund Freud, in contrast, proposed that infantile amnesia was a form of repression—in other words, a defense mechanism against disagreeable or negative recollections.

Researchers have concluded that the infant brain loses memories far more quickly than does the developed brain and that it lacks the ability to generalize to new events.

Children under the age of five or six do not yet realize that learning is most effective when new information is associated meaningfully with other knowledge.

Young children are similarly unaware that the intentional rehearsal or repetition of new information will enhance their ability to retain it in memory.

As children age and develop language expertise, however, they begin to draw upon their semantic memory to help them remember words, facts, and events.

They also grow increasingly aware of the ways in which memory serves them. This awareness of how memory works, known as "metamemory," increases through much of adulthood.

Aging

Older adults experience memory loss, but only for memories of certain types. Episodic memory (the ability to remember specific events) is typically the first type of memory to decline in old age; it is also the last to fully develop in children.

Associative memory (the ability to learn, store, and retrieve associations between actions or things) also declines dramatically.

ART FOR PSYCHOTHERAPY

In fact, a chief memory complaint among older adults is a decreasing ability to associate a person's name with his face.

Studies conducted separately by American psychologists Marcia K. Johnson and Larry L. Jacoby demonstrated that, whereas older adults are able to remember the gist of an action or event just as well as younger adults, they are unable to recollect the specific details that were involved.

Older adults also have particular difficulty remembering the source of their memories, even in cases in which the information is familiar. Yet other types of memory are spared in old age—the most common among these being recognition.

It is therefore common for an older adult to recognize a person's face while failing to recall that person's name.

Jacoby's work measured age-dependent distinctions between familiarity (recognition) and source memory (recollection) of a given event. His studies provided stronger confirmation that recognition abilities are similar in younger and older individuals, but as people age, they are less able to recall specific details of the events related to the familiar person or thing.

Changes in the brains of older adults, especially in the frontal lobes and hippocampal area, also may result in age-related memory deficits. More severe and widespread changes in the brain are related to the massive declines in memory functioning seen in Alzheimer disease, also known as Senile Dementia of Alzheimer Type or SDAT.

ART FOR PSYCHOTHERAPY

Causes of Amnesia

Loss of memory occurring most often as a result of damage to the brain from trauma, stroke, Alzheimer disease, alcohol and drug toxicity, or infection.

Amnesia may be anterograde, in which events following the causative trauma or disease are forgotten, or retrograde, in which events preceding the causative event are forgotten.

The condition also may be traced to severe emotional shock, in which case personal memories (e.g., identity) are affected. Such amnesia seems to represent a psychological escape from or denial of memories that might cause anxiety.

These memories are not actually lost, since they can generally be recovered through psychotherapy or after the amnesic state has ended.

Occasionally amnesia may last for weeks, months, or even years, during which time the person may begin an entirely new life. Such protracted reactions are called fugue states.

When recovered, the person is usually able to remember events that occurred prior to onset, but events of the fugue period are forgotten.

Posthypnotic amnesia, the forgetting of most or all events that occur while under hypnosis in response to a suggestion by the hypnotist, has long been regarded as a sign of deep hypnosis.

The common difficulty of remembering childhood experiences is sometimes referred to as childhood amnesia.

ART FOR PSYCHOTHERAPY

SECTION EIGHT: SORCERY OR MAGIC

Magic

This is a concept used to describe a mode of rationality or way of thinking that looks to invisible forces to influence events, effect change in material conditions, or present the illusion of change.

Within the Western tradition, this way of thinking is distinct from religious or scientific modes; however, such distinctions and even the definition of magic are subject to wide debate.

Nature and scope

Practices classified as magic include divination, astrology, incantations, alchemy, sorcery, spirit mediation, and necromancy. The term *magic* is also used colloquially in Western popular culture to refer to acts of conjuring and sleight of hand for entertainment.

The purpose of magic is to acquire knowledge, power, love, or wealth; to heal or ward off illness or danger; to guarantee productivity or success in an endeavour; to cause harm to an enemy; to reveal information; to induce spiritual transformation; to trick; or to entertain.

The effectiveness of magic is often determined by the condition and performance of the magician, who is thought to have access to unseen forces and special knowledge of the appropriate words and actions to manipulate those forces.

ART FOR PSYCHOTHERAPY

Phenomena associated or confused with magic include forms of mysticism, medicine, paganism, heresy, witchcraft, shamanism, voodoo, and superstition.

Magic is sometimes divided into the "high" magic of the intellectual elite, bordering on science, and the "low" magic of common folk practices. A distinction is also made between "black" magic, used for nefarious purposes, and "white" magic, ostensibly used for beneficial purposes.

Although these boundaries are often unclear, magical practices have a sense of "otherness" because of the supernatural power that is believed to be channelled through the practitioner, who is a marginalized or stigmatized figure in some societies and a central one in others.

Elements of magic

The performance of magic involves words (e.g, spells, incantations, or charms) and symbolic numbers that are thought to have innate power, natural or man-made material objects, and ritual actions performed by the magician or other participants.

A spell or incantation is believed to draw power from spiritual agencies to accomplish magic. Knowledge of spells or symbolic numbers is often secret (occult), and the possessor of such knowledge can be either greatly revered or feared.

In some cases, the spell is the most highly regarded component of the magical rite or ceremony.

ART FOR PSYCHOTHERAPY

The Trobriand Islanders of Melanesia, for example, regarded using the right words in the right way as essential to the efficacy of the rite being performed.

Among the Maori of New Zealand the power of words is thought to be so important that mistakes in public recitations are believed to cause disasters for individuals or the community.

Moreover, like the medieval European charms that used archaic languages and parts of the Latin liturgy, spells often employ an esoteric vocabulary that adds to the respect accorded rites.

Belief in the transformative power of words is also common in many religions. Shamans, spirit mediums, and mystics, for example, repeat specific sounds or syllables to achieve an ecstatic state of contact with spiritual forces or an enlightened state of consciousness.

Even modern magic for entertainment retains a residual of the spell with its use of the term *abracadabra*.

Material

Much anthropological literature refers to the objects used in magic as "medicines," hence the popular use of the term *medicine man* for magician.

These medicines include herbs, animal parts, gemstones, sacred objects, or props used in performance and are thought to be potent in themselves or empowered by incantations or rituals.

In some cases, medicines that are intended to heal are physiologically effective; for example, the poppy is used widely as an anesthetic, willow bark is used by some

ART FOR PSYCHOTHERAPY

Chinese as an analgesic, and garlic and onions were used as antibiotics in medieval Europe.

Other medicines that are meant to cause harm, such as toad extracts and bufadienolides, are, in fact, known poisons. Other materials have a symbolic relationship to the intended outcome, as with divination from animal parts.

In scapulamancy (divination from a sheep shoulder bone), for example, the sheep's bone reflects the macrocosmic forces of the universe.

In sorcery, a magician may employ something belonging to the intended victim (e.g., hair, nail parings, or a piece of clothing) as part of the ritual.

The rite itself may be symbolic, as with the drawing of protective circles in which to call up spirits, the sprinkling of water on the ground to make rain, or the destruction of a wax image to harm a victim.

Plants or other objects can also symbolize desired outcomes. In rites to ensure a canoe's speed, the Trobriand use light vegetable leaves to represent the ease with which the craft will glide over the water; the Zande of South Sudan place a stone in a tree fork to postpone the setting of the sun; and many Balkan peoples once swallowed gold to cure jaundice.

Rites and condition of the performer

Because magic is based on performance, ritual and the magician's knowledge and ability play a significant role in its efficacy.

ART FOR PSYCHOTHERAPY

The performance of magic also presumes an audience, either the spiritual forces addressed, the patient-client, or the community. Both the magician and the rite itself are concerned with the observance of taboos and the purification of the participants.

Magicians, like priests presiding over religious rituals, observe restrictions of diet or sexual activity to demarcate the rite from ordinary and profane activities and to invest it with sanctity.

Modern magicians' success with entertaining audiences is dependant primarily on their performance skills in manipulating material objects to create an illusion.

Functions

Foremost among the many roles magic plays are its "instrumental" and "expressive" functions. Based in the attempt to influence nature or human behaviour, magic's instrumental function is measured by its efficacy in achieving the desired result.

Anthropologists identify three main types of instrumental magic: the productive, the protective, and the destructive. Productive magic is employed to solicit a successful outcome from human labour or nature, such as bountiful hunt or harvest or good weather.

Protective magic aims to defend an individual or community from the vagaries of nature and the evil of others. The use of amulets to ward off contagious diseases or the recitation of charms before a journey are examples of this protective function.

Lastly, destructive magic, or sorcery, is intended to harm others, often is motivated by envy, and is socially disruptive. Consequently, the use of counter-magic against sorcery may relieve some social tension within a community.

Magic's expressive function results from the symbolic and social meanings attached to its practices, though its performers may not necessarily be aware of this function.

Magic can provide a sense of group identity through shared rituals that give power or strength to members. At the same time, it can isolate the magician as a special person within or on the margins of society.

Magic can also serve as a creative outlet or form of entertainment. It is, therefore, inseparable from the total system of thought, belief, and practice in a given society.

Definitional issues: magic, religion, and science

The term *magic* cannot be defined in isolation because of its broad parameters, important role in many societies, and interactions with related phenomena.

Magic is a generic label used by outsiders (theoretically, objective observers) to describe specific practices in societies in which this word or its conceptual equivalent may not even exist.

As a result, diverse phenomena are lumped together on the assumption that they operate in the same way. This artificial construct of magic also exists only in relation to what it is not—primarily, religion and science as alternate modes of rationality.

Such definitions of magic privilege cultures with a strong scientific orientation and stigmatize those that practice magic instead of religion. Consequently, defining magic and identifying magicians requires an understanding of the cultural contexts in which these labels are used.

Although magic has an ambiguous relationship with Western religion and science, it is rooted in the main institutional, social, and intellectual traditions in Western history. Moreover, modern attempts to arrive at a universal definition of magic reflect a Western bias.

In particular, 18th- and 19th-century views on cultural and historical evolution set magic apart from religion and science. In a model developed by the British anthropologist Sir James Frazer (1854–1941), magic is characterized as an early stage in human development, superseded first by religion and then by science.

The debate over the relationship between magic, religion, and science that dominated much of the discussion about magic throughout the 20th century is evident in the fieldwork of anthropologists, the theories of sociologists of religion, and critiques by postmodernists.

Consequently, research in comparative religions, history, and anthropology in the second half of the 20th century moved away from the evolutionary model toward more context-sensitive interpretations, while other studies developed new models for cross-cultural comparison.

Nonetheless, the magic-religion-science model retains considerable interpretive power, and the dichotomies used to distinguish magic from religion or science are pervasive in popular discourse.

Magic and religion

Magic continues to be widely perceived as an archaic worldview, a form of superstition lacking the intrinsic

ART FOR PSYCHOTHERAPY

spiritual value of religion or the rational logic of science. Religion, according to seminal anthropologist Sir Edward Burnett Tylor (1832–1917), involves a direct, personal relationship between humans and spiritual forces; in religion's highest form, that relationship is with a personal, conscious omnipotent spiritual being.

Magic, on the other hand, is characterized as external, impersonal, and mechanical, involving technical acts of power.

Magic seeks to manipulate spiritual powers, while religious prayer supplicates spiritual forces; a distinction explored by Bronisław Malinowski (1884–1942) in his work on the Trobriand Islanders. Moreover, according to Émile Durkheim (1858–1917), religion is communal because its adherents, bound together by shared belief, form a church.

Magic, on the other hand, involves no permanent ties between believers and only temporary ties between individuals and the magicians who perform services for them. The fieldwork of A.R. Radcliffe-Brown (1881–1955) among the Andaman Islanders, however, has made clear that magic, too, may have a communal dimension.

Magic and science

Although magic is similar in some respects to science and technology, it approaches efficacy (the ability to produce a desired material outcome) differently.

Magic, like religion, is concerned with invisible, non-empirical forces; yet, like science, it also makes claims to efficacy.

ART FOR PSYCHOTHERAPY

Unlike science, which measures outcomes through empirical and experimental means, magic invokes a symbolic cause-effect relationship.

Moreover, like religion and unlike science, magic has an expressive function in addition to its instrumental function.

Magical rainmaking strategies, for example, may or may not be efficacious, but they serve the expressive purpose of reinforcing the social importance of rain and farming to a community.

Subcategories of magic

The view of magic as pre-religious or non-scientific has contributed both to subtle distinctions between magic and other practices and to the recognition of subcategories of magic.

Notably, anthropologists distinguish magic from witchcraft, defining the former as the manipulation of an external power by mechanical or behavioural means to affect others and the latter as an inherent personal quality that allows the witch to achieve the same ends.

However, the line between the two is not always clear, and in some parts of the world an individual may operate in both ways. Similarly, the distinction between "black" magic and "white" magic is obscure since both practices often use the same means and are performed by the same person.

Scholars also distinguish between magic and divination, whose purpose is not to influence events but to predict or understand them.

ART FOR PSYCHOTHERAPY

Nevertheless, the mystical power of diviners may be thought to be the same as that behind magic.

Ultimately, despite these distinctions and the variety of unique roles that practitioners play in their own societies, most end up classified under the universal term *magician*.

Often even religious figures such as priests, shamans, and prophets are identified as magicians because many of their activities include acts defined as "magical" by modern scholars.

In the end, distinctions between magic and religion or science are harder to make in practice than in theory; scholars therefore use labels such as *magico-religious* to describe activities or persons who cross this artificial dividing line.

Similarly, the boundary between magic and science is permeable, since the modern scientific method (observation and experimentation) evolved from forms of scientific magic such as alchemy and astrology.

Thus, the evolutionary model, which draws sharp distinctions between magic, religion, and science, cannot account for the essential similarity between various phenomena.

Moreover, dichotomies that define magic in relation to other phenomena are reductionist, often ignoring the meaningful structures and beliefs that inform these practices in their native context.

Conceptual history

The claim that magic is found in all human societies rests on a definition that is rooted in Western cultural assumptions, and both these assumptions and the use of the term *magic* have undergone change over time and place.

Consequently, to understand beliefs and practices in other societies that appear similar to European magic, it is necessary to apply the context-sensitive and comparative methods that become increasingly important in the study of anthropology, history, and religion.

History of magic in Western worldviews

The Western conception of magic is rooted in the ancient Judeo-Christian and Greco-Roman heritage. The tradition took further shape in northern Europe during the medieval and early modern period before spreading to other parts of the globe through European exploration and colonialism after 1500.

The view of Western civilization as a story of progress includes the magic-religion-science paradigm that traces the "rise" and "decline" of magic and then religion, along with the final triumph of science—a model now challenged by scholars. Moreover, the very origins of the word *magic* raise questions about ways in which one person's religion is another person's magic, and vice versa.

Ancient Mediterranean world

The root word for *magic* (Greek: *mageia*; Latin: *magia*) derives from the Greek term *magoi*, which refers to a Median tribe in Persia and their religion, Zoroastrianism.

The Greco-Roman tradition held that magicians possessed arcane or secret knowledge and the ability to

ART FOR PSYCHOTHERAPY

channel power from or through any of the polytheistic deities, spirits, or ancestors of the ancient pantheons.

Indeed, many of the traditions associated with magic in the Classical world derive from a fascination with ancient Middle Eastern beliefs and are concerned with a need for counter-magic against sorcery.

Spells uttered by sorcerers and addressed to gods, to fire, to salt, and to grain are recorded from Mesopotamia and Egypt.

These texts also reveal the practice of necromancy, invoking the spirits of the dead, who were regarded as the last defence against evil magic. Greco-Egyptian papyruses from the 1st to the 4th century CE, for example, include magical recipes involving animals and animal substances, along with instructions for the ritual preparations necessary to ensure the efficacy of the spells.

Divination took many forms—from the Etruscan art of haruspicina (reading entrails of animal sacrifices) to the Roman practice of augury (interpreting the behaviour of birds)—and was widely practiced as a means of determining propitious times to engage in specific activities; it often played a role in political decision making. Ancient Roman society was particularly concerned with sorcery and counter-sorcery, contests associated with the development of competitive new urban classes whose members had to rely on their own efforts in both material and magical terms to defeat their rivals and attain success.

Ambivalence toward magic carried into the early Christian era of the Roman Empire and its subsequent heirs in Europe and Byzantium.

ART FOR PSYCHOTHERAPY

In the Gospel According to Matthew, the Magi who appeared at the birth of Jesus Christ were both Persian foreigners of Greco-Roman conception and wise astrologers. As practitioners of a foreign religion, they seemed to validate the significance of Jesus' birth.

However, *magus*, the singular form of *magi*, has a negative connotation in the New Testament in the account of Simon Magus (Acts 8:9–25), the magician who attempted to buy the miraculous power of the Disciples of Christ.

In medieval European Christian legends, his story developed into a dramatic contest between true religion, with its divine miracles, and false demonic magic, with its illusions.

Nonetheless, belief in the reality of occult powers and the need for Christian counter-rituals persisted, for example, in the Byzantine belief in the "evil eye" cast by the envious, which was thought to be demonically inspired and from which Christians needed protection through divine remedies.

Medieval Europe

During the period of Europe's conversion to Christianity (*c.* 300–1050), magic was strongly identified with paganism, the label Christian missionaries used to demonize the religious beliefs of Celtic, Germanic, and Scandinavian peoples.

Church leaders simultaneously appropriated and Christianized native practices and beliefs. For example, medicinal remedies found in monastic manuscripts combined Christian formulas and rites with Germanic folk rituals to empower natural ingredients to cure ailments caused by poisons, elf-attack, demonic possession, or other invisible forces.

Another Christianized practice, bibliomancy (divination through the random selection of a biblical text), was codified in the 11th-century *Divinatory Psalter* of the Orthodox Slavs. Although co-opted and condemned by Christian leaders of this period, magic survived in a complex relationship with the dominant religion.

Similar acculturation processes occurred in later conversions in Latin America and Africa, where indigenous beliefs in spiritual forces and magical practices coexist, sometimes uneasily, with Christian theology.

In high medieval Europe (*c.* 1050–1350), the battle between religion and magic occurred as the struggle against heresy, the church's label for perverted Christian belief. Magicians, like heretics, were believed to distort or abuse Christian rites to do the Devil's work.

By the 15th century, belief in the reality of human pacts with the Devil and the magical powers acquired through them contributed to the persecution of those accused of actually harming others with their magic. Also in the high Middle Ages the demonization of Muslims and Jews contributed to the suspicion of the "other."

Marginal groups were routinely accused of ritual baby killing. In lurid accounts of the "blood libel," Jews were charged with stealing Christian children for sacrifice.

The ancient Romans made similar accusations against witches by Christians and against Christians.

ART FOR PSYCHOTHERAPY

Although magic was widely condemned during the Middle Ages, often for political or social reasons, the proliferation of magic formulas and books from the period indicates its widespread practice in various forms.

Richard Kieckhefer has identified two major categories of magic: "low" magic includes charms (prayers, blessings, adjurations), protective amulets and talismans, sorcery (the misuse of medical and protective magic), divination and popular astrology, trickery, and medical magic through herbs and animals; and "high," or intellectual, magic, includes more learned forms of astrology, astral magic, alchemy, books of secrets, and necromancy.

There is also evidence of courtly interest in magic, particularly that involving automatons and gemstones. Moreover, magic served as a literary device of the time, notably the presence of Merlin in the Arthurian romances.

Although medieval European magic retained its sense of otherness by borrowing from Jewish practices and Arabic scientific sources such as the astral magic manual *Picatrix*, it also drew from the mainstream Christian tradition. Necromancy, for example, used Latin Christian rites and formulas to compel the spirits of the dead to obey.

Late medieval and early modern Europe

By the late Middle Ages (c. 1350–1450) and into the early modern period (c. 1450–1750), magic was regarded as part of a widespread and dangerously antisocial demonic cult that included the condemned practices of sorcery, necromancy, and witchcraft.

ART FOR PSYCHOTHERAPY

Accused heretics, witches, and magicians were subject to inquisitions designed to uncover these cult connections and to destroy the means of transmission (e.g., the burning of condemned books and/or the "guilty" parties).

The influential manual *Malleus maleficarum* ("The Hammer of Witches," 1486) by Jacob Sprenger and Henry Krämer describes witchcraft in great detail (e.g, the witches' sabbath, a midnight assembly in fealty to the Devil); moreover, this oft-reprinted volume is responsible for the misogynist association of witchcraft with women that becomes the dominant characteristic in the early modern period.

This conspiracy theory of demonic magic contributed to the early modern "witch craze" that occurred at a time of growing tension between magic, religion, and nascent science.

Nonetheless, despite the persecution of "black" magic and its alleged practitioners, forms of "white" magic persisted in Europe on the boundaries between magic, mysticism, and emerging empiricism.

During the Renaissance, there was renewed interest in ancient Middle Eastern practices, Neoplatonic mysticism, and Arabic texts on alchemy and astrology. Pico della Mirandola sought hidden knowledge in Jewish Kabbala, a mystical practice for unlocking the divine secrets contained in written and unwritten Hebrew Scriptures.

Marsilio Ficino studied astral magic and the power of music to channel cosmic influences, while Giordano Bruno explored the mystical traditions of Hermeticism, based on works of the legendary Alexandrian prophet of the 1st–3rd century Hermes Trismegistus.

ART FOR PSYCHOTHERAPY

Although generally tolerated because their practices were perceived to be within the main Judaic and Christian Hermetic tradition, practitioners of alchemy were sometimes considered to be evil magicians who acquired their knowledge through a pact with the Devil (as in the Faust legends).

When magical activities of intellectual dilettantes proved, or appeared, to be antisocial, the results were more often put down to simple trickery—as in the case of the 18th-century charlatan Alessandro, conte di Cagliostro (Giuseppe Balsamo).

European traditions and the modern world

The European fascination with the magical traditions of the ancient Middle East was extended to those of East and South Asia when Europeans made contact with these regions in the early modern period.

Orientalism, as literary and cultural critic Edward Said labelled this phenomenon, has its roots in the sense of the "other" found in the earliest definitions of magic (notably the Magi as Persian foreigners) and in the Renaissance penchant for Egyptian, Hebrew, and Arabic materials.

Intrigued by the exotic otherness of Eastern societies, modern European philosophers experimented with the progressive model of magic-science-religion. Georg Wilhelm Friedrich Hegel, for example, viewed 19th-century India as an immature civilization, in part because Hindu consciousness lacked the categories of logic Hegel valued.

ART FOR PSYCHOTHERAPY

A popular "scientific" worldview prevails in modern Western societies that suggest the triumph of human reason.

Enlightenment rationalism and the scientific revolution—ironically rooted in Renaissance experiments in magic and motivated in part by Reformation pragmatism—led to the modern triumph of scientific reasoning over magic, evident, for example, in 19th-century exposés of magic tricksters as charlatans.

Notably, spirit rappers, mediums who "conversed" with spirits who replied by knocking on a table, were easily exposed as the ones doing the knocking.

Modern popular magic has appeared in the realm of entertainment, generally as a plot device in stories and movies, as tricks aimed at children, and as mysterious sleight-of-hand illusions in magic shows that delight the audience's sense perceptions and challenge their reasoning ability.

The fascination with occult knowledge and mystical powers derived from non-mainstream or foreign sources persists in the West in astrological charts in newspapers, theories of interplanetary aliens and government conspiracies to hide them,.

Occult rituals in some New Age religions, and interest in traditional practices that have an esoteric flavour, such as *feng shui* (geomancy, the traditional Asian practice of aligning graves, homes, and temples with cosmic forces).

This persistence suggests, in part, the impact of globalization on postmodern worldviews challenging the dominance of a strictly scientific mode of rationality.

ART FOR PSYCHOTHERAPY

Globalization of the magic concept

Western conceptions of magic, religion, and science were exported to other parts of the globe in the modern period by traders, conquerors, missionaries, anthropologists, and historians.

European travellers in the 16th–19th centuries functioned as primitive ethnographers whose written observations are invaluable historical resources. However, their accounts, often coloured by their Judeo-Christian assumptions about religion versus magic, illuminate how indigenous peoples were treated as "children" to be educated or, in the case of some conquerors, as subhuman races to be enslaved.

During the latter part of the 19th century, anthropologists began to analyze magic and its part in the evolution of the world's religions.

Their work was characterized by a fundamental distinction rooted in the magic-religion-science evolutionary model: the world is divided between historical, literate urbanized cultures, or "civilizations" (for example, the ancient traditions of East and South Asia) and non-literate, tribal archaic, or "primitive," societies (such as those found in parts of Africa, the Americas, and Oceania). Historians viewed complex societies characterized by urbanization, centralization, and written traditions as more advanced and measured their progress as civilizations according to the evolutionary model.

Early European observers as developmentally stagnant people without history often perceived nomadic, tribal, agricultural, or non-urbanized societies with strong oral traditions.

ART FOR PSYCHOTHERAPY

While these views are no longer accepted, their residual effect is still felt in the way magic, religion, and science are conceptualized.

Anthropologists of religion traditionally distinguished between the "religion" practiced by the world's main faiths, which often marginalize magic as superstition, and the beliefs of small non-literate societies in which "magic" may in fact be central to religious belief.

Here the distinction between religion and magic seems unfounded. Indeed, as some postcolonial societies endeavour to distance themselves from Western logic, ancient religious traditions are pivotal to the reassertion of cultural identity and autonomy.

West African vodun (voodoo), which spread to the Caribbean, the Americas, and elsewhere, is one example of an indigenous religious practice that is tied to cultural identity in art, music, and literature and used subversively as a rallying point for postcolonial resistance to Western modes of rationality.

World cultures

The Western concept of magic as a set of beliefs, values, and practices that are not fully religious or scientific does not find its equivalent in non-Western languages and cultures; conversely, concepts found in other cultures may be untranslatable into English or a Western framework.

For example, Hawaiian historian David Malo (c. 1793–1853), discussing Christianity and traditional Hawaiian religion, found *ho`omana* (to make, to do, or to imbue with supernatural, divine, or miraculous power) the closest translation for English *religion*, contrary to its characterization by Westerners as a magical component in Polynesian beliefs.

Furthermore, a modern Japanese dictionary uses a transliteration, *majikku*, for the English word *magic*. It also uses the English word *magic* to translate several Japanese words beginning with *ma-*, the kanji character representing a vengeful spirit of the dead (in East Asian folk belief, an ancestor not cared for properly; in Buddhist cosmology, an evil demonic figure).

While superficially similar to the Christian notion of magic as demonic, the cosmologies regarding these demons differ significantly. Moreover, *ma-* does not have the range of meanings that *magic* has in Western thought.

On the other hand, specific practices identified as magic—e.g., divination, spells, spirit mediation—are found worldwide, even if the word *magic* is not.

For example, in China various practices such as divination through oracle bones, offerings to dead ancestors, and *feng shui* can be classified as either magic, religion, or science.

But, it is questionable whether these categories have any validity in Chinese thought; rather these so-called magical practices are an intrinsic part of the worldviews expressed in China's main religious and philosophical systems (ancestor worship, Confucianism, Daoism, and Buddhism).

In modern China, some communities deal with crisis by combining seemingly contradictory practices—including supplication and coercion of gods, appeals to ancestral spirits, folk cures, and modern inoculations. Such

ART FOR PSYCHOTHERAPY

syncretism has been common in East Asia; notably, in 6th-century Japan the native nature worship of Shinto blended with imported forms of Buddhism without the kind of conflict that occurred during the conversion of Europe to Christianity.

In modern East Asia, conflict between magic, religion, and science introduced by Western concepts of magic occurs alongside a strong tradition of syncretism that blends empirical science with practices that Westerners often perceive as unscientific magic or religious superstition.

Asian religious traditions such as Hinduism, Buddhism, and Daoism teach that material life is illusory. This mode of rationality focuses on understanding the principles and spiritual forces that lie behind physical experience.

Consequently, adepts in these traditions who have achieved a level of understanding of these cosmic forces often appear to have the ability to manipulate physical reality in ways that seem magical.

The point of demonstrations by street magicians and snake charmers in India is to show the illusory quality of material reality in order to draw attention to the universal, timeless, and cosmic.

Purposeful deception in magic is thus used to illustrate the deceptiveness of human apprehensions of reality. The mystical component of magic is also clear in Tantra and other esoteric and nonconformist sects of Hinduism or Buddhism, which use mystical words, symbols, and diagrams in their rituals.

Whether these practices are magic or religion depends upon one's point of view.

ART FOR PSYCHOTHERAPY

Post-colonial points of views

Anthropological and sociological studies of modern non-literate societies in the Americas, Oceania, and Africa have given rise to new global terminology.

Beginning in the second half of the 20th century, some sociologists and anthropologists turned the tables on earlier scholarship by applying the methods used for examining extant non-literate ("primitive") societies to literate, urban societies of the past, which previously had been evaluated by the criteria reserved for the study of "civilizations."

For example, the phenomenon of shamanism and the word *shaman*, as defined by Mircea Eliade (1907–86) in his exploration of ecstatic states, has been applied not only to "primitive" cultures but to premodern Christian Europe.

Likewise the term *mana* ("power"), appropriated from Melanesian and Polynesian cultures by Émile Durkheim and Marcel Mauss (1872–1950), has been widely applied to magical practices in historical civilizations, including that of Classical Rome.

History of magic theories

Because of the impact of anthropological theory on the study of magic, its development and history bear reviewing.

The first important figure in this line of inquiry was Sir Edward Burnett Tylor, whose *Primitive Culture* (1871) regarded magic as a "pseudo-science" in which the

ART FOR PSYCHOTHERAPY

"savage" postulated a direct cause-effect relationship between the magical act and the desired outcome.

Tylor regarded magic as "one of the most pernicious delusions that ever vexed mankind," but he did not approach it as superstition or heresy. Instead he studied it as a phenomenon based on the "symbolic principle of magic," a scheme of thought founded on a rational process of analogy.

He also realized that magic and religion are parts of a total system of thought. Although he believed that magic and animistic beliefs became less prevalent in the later stages of history, he did not view magic and religion as alternative stages in the evolutionary development of humankind.

That conclusion would be left for Sir James Frazer in *The Golden Bough* (1890), in which he ordered magic, religion, and science in a grandiose evolutionary scheme. Magic preceded religion because, according to Frazer, the former was logically simpler.

This notion, however, was a based on his erroneous assumption that the Australian Aborigines, examples of a "primitive" people, believed in magic but not in religion.

Sociological theories

Another line of theorists, including sociologists Durkheim and Mauss, widened the discussion by defining magic in terms of its social function.

In *The Elementary Forms of the Religious Life* (1912), Durkheim argued that magical rites involved the manipulation of sacred objects by the magician on behalf of individual clients; the socially cohesive significance of religious rites proper (by priests) was therefore largely lacking.

Durkheim's views were furthered by A.R. Radcliffe-Brown in the *The Andaman Islanders* (1922) and to a lesser extent by Malinowski in *Argonauts of the Western Pacific* (1922) and *Magic, Science and Religion* (1925). Radcliffe-Brown posited that the function of magic was to express the social importance of the desired event, while Malinowski regarded magic as directly and essentially concerned with the psychological needs of the individual.

Subsequent studies of the working of systems of magic, especially in Africa and Oceania, built upon the work of Malinowski and Radcliffe-Brown along with that of Sir Edward Evans-Pritchard in *Witchcraft, Oracles and Magic Among the Azande* (1937). In his seminal book, Evans-Pritchard demonstrated that magic is an integral part of religion and culture used to explain events that cannot otherwise be understood or controlled.

The Zande of South Sudan accept magic, together with witchcraft and oracles, as a normal part of nature and society. These phenomena form a closed logical system, each part of which buttresses the other and provides a rational system of causation.

Psychological theories

These anthropological and sociological approaches focused on magic as a social phenomenon, but the role of individual psychology was implicit in the views of Tylor and Frazer and brought out more in the work of Malinowski, who frequently offered psychological explanations for belief in magic.

ART FOR PSYCHOTHERAPY

Sigmund Freud's influential view of magic as the earliest phase in the development of religious thought (*Totem and Taboo*, 1918) followed Frazer's model and posited an essential similarity between the thought of children, neurotics, and "savages."

According to Freud, all three assumed that wish or intention led automatically to the fulfilment of the desired end.

This reductionist view, based on outmoded notions about "primitive" cultures, was revised as the result of new field research. Although Claude Lévi-Strauss also initially equated these three groups, he later modified this view in his analysis of the work of Mauss, which focuses on the structural linguistics of terms such as *mana* that are deployed in the study of magic.

His work, therefore, laid the foundation for later deconstructions of the concept of magic.

Comparative religions

The rise of the study of comparative religion led to new theories that accounted for both world religions and localized belief systems.

The work of Eliade, including his study of shamanism, is an important and influential example of this approach, as is that of Ninian Smart, who devised a six-dimensional (experiential, mythic, doctrinal, ethical, ritual, and social) worldview analysis for cross-cultural comparison that can be applied to different belief systems, whether called magic or religion.

Likewise, Judaic scholar Jacob Neusner suggested the neutral rubric "modes of rationality" to avoid pejorative

ART FOR PSYCHOTHERAPY

comparisons between systems of thought otherwise classified as magic, religion, science, or philosophy. T

he broader base established by the comparative religions approach avoids the difficulties of distinguishing urban literate from non-urban non-literate societies and the perils of the magic-religion-science progression.

Post-modern dialogue

Postmodern scholarship continues to challenge older anthropological notions. The work of such anthropologists as Victor Turner (1920–83), Clifford Geertz, and Marshall Sahlins has had a wide impact on the social sciences and humanities.

Central to the challenge to the traditional magic-religion-science paradigm was *Magic, Science, Religion, and the Scope of Rationality* (1990), in which Stanley Jeyaraja Tambiah deconstructs the European history of the progress model and the work of anthropologists from Tylor forward.

Other anthropologists have questioned the model of the rise and decline of magic in European thought articulated in Keith Thomas's groundbreaking *Religion and the Decline of Magic* (1971), a study of early modern England, and Valerie Flint's *The Rise of Magic in Early Medieval Europe* (1991).

These cross-disciplinary debates, along with the rejection of the Western magic-religion-science paradigm, have contributed to more sensitive treatments of magical practices in diverse societies.

ART FOR PSYCHOTHERAPY

Sorcery

The practice of malevolent magic, derived from casting lots as a means of divining the future in the ancient Mediterranean world.

Some scholars distinguish sorcery from witchcraft by noting that it is learned rather than intrinsic.

Other scholars, noting that modern witches claim to learn their craft, suggest that sorcery's intent is always evil and that of witchcraft can be either good or bad. In the early Christian era, the term was applied to any magician or wizard but by the Middle Ages only to those who allegedly practiced magic intended to harm others.

In Western popular culture and in Western children's literature in particular, the sorcerer often assumes a more positive guise.

The sorcerer has traditionally been feared, in part because of his supposed knowledge of the occult and especially because of his understanding of poisons.

Indeed, in the 13th and 14th centuries, most trials for "witchcraft" involved deaths attributed to malevolent magic but which were probably caused by poisoning. In a famous case in 1324 in Ireland, Lady Alice Kyteller was charged with performing magical rites, having sexual intercourse with demons, attempting to divine the future, and poisoning her first three husbands.

In the *Malleus Maleficarum* (1486, "The Hammer of Witches"), the famous witch-hunter's manual, Dominicans Heinrich Krämer and Jacob Sprenger associated the practice of sorcery with a group of "witches" who allegedly practiced Satanism.

ART FOR PSYCHOTHERAPY

Because of their work and that of others in the 13th–15th centuries, witchcraft was understood as a Christian heresy, and sorcery, like the practice of magic in general, was believed to be an integral part of a witch's dealings with the Devil.

In the early modern period, those who were known to pronounce curses were guilty of sorcery.

Notably, the witchcraft trials in Salem, Massachusetts, in the 17th century were rooted in accusations against two women who had allegedly cursed their neighbour's cows and caused them to stop producing milk.

Contemporary witches, or Wiccans, do not practice Satanism and have denounced the practice of malevolent magic.

Conclusion on magic and sorcery

The study of magic and sorcery as a distinct cultural phenomenon has a long history in anthropological, sociological, and historical studies.

Although some distinctions between magic and other religious or scientific activities may be useful, magic cannot be studied in isolation as it once was.

Practices classified as magic represent essentially an aspect or reflection of the worldview held by a particular people at a particular point in their own historical development.

Magic and sorcery, like religion and science, is thus a part of a culture's total worldview.

ART FOR PSYCHOTHERAPY

SECTION NINE: PROFESSIONAL ART THERAPY

Skilled communication for therapy

Art therapy is a valuable therapy with which to release past trauma and recondition established habits.

Even though our personal unconscious only ever seeks to promote our well being it can often be the seat of faulty learning from our childhood, leading to low self esteem, under achievement and sometimes worse.

Often it attempts to protect us by raising our fears and anxieties to phobic levels to keep us from a particular activity or stimulus it sees as dangerous. Utilising art in therapy often facilitates an unconscious relearning process.

An artistic projection occurs normally in everyone when certain physiological and psychological conditions are met and with the assistance of a skilled art therapist, it is possible to use this state to make deep and lasting changes to thoughts, feelings and behaviour

Art therapy is a skilled communication aimed at directing a person's imagination in a way that helps elicit changes in some perceptions, sensations, feelings, thoughts and behaviours. In a typical art therapeutic session, the therapist and client will discuss the intended alterations or therapeutic goals desired.

The therapist will ask questions about previous medical history, general health, and lifestyle to decide on the best approach for the individual.

Art therapy may be found to be helpful for those seeking relief from a range of problems and is used alongside a

ART FOR PSYCHOTHERAPY

person's own willpower and motivation to seek a desired goal.

Choosing a practitioner is important to choose a qualified therapist who has undertaken all the necessary training to understand the theory and practice of art therapy.

State of consciousness

Art is projected in a different state of consciousness, which you can naturally enter so that, for therapeutic purposes, beneficial corrections may be given directly to your unconscious mind.

In this way, art therapy is an effective way of making contact with our inner (unconscious) self, which is both a reservoir of unrecognised potential and knowledge as well as being the unwitting source of many of our problems.

Realistically no one can be influenced against their will and even when in session, a person can still reject any suggestion. Thus, art therapy is a state of purposeful co-operation.

Art therapy makes use of the bicameral nature of the functioning brain and the conscious/ unconscious processes therein.

At its simplest level, the unconscious mind becomes (through life experience) the repository of conditioned experience, while the conscious mind is the waking mind dealing with appraisal and decision making.

Art therapy is used to access and re-programme complexes, which are being sustained and remain active at the unconscious level.

ART FOR PSYCHOTHERAPY

Formulation of a session

The art therapist will ask about the problem. The length of treatment will be discussed (whether one session or several) and the fee will be agreed.

The therapist will take a 'case history' of the circumstances and history of the particular problem and help with the formulation of an efficient treatment strategy.

There may be some preliminary discussion and explanation of the principles of art therapy before starting the full treatment, either immediately or in a subsequent session.

One of the great things about art therapy is that in the hands of a reputable practitioner it is harmless and produces no damaging side effects, whilst providing positive therapeutic benefits.

Art therapy is a two way process between the therapist and the client, a professional partnership.

Ethical approach

Trained and long serving professional art therapists consider their ethical approach as being as valuable as psychotherapy.

As in psychotherapy, the treatment of emotional or behavioural problems is by psychological means, often in one-to-one interviews or small groups.

Freud devised the first systematic approach, initially discussing patients' problems with them, but later

ART FOR PSYCHOTHERAPY

allowing them to do most of the talking in a procedure called free association of ideas.

This has been the model for subsequent psychotherapies and hypnotherapeutic methods; modern psychoanalysis and cognitive therapies associated with theories such as learned helplessness concentrate on the patient's beliefs.

Other therapies, such as those within humanistic psychology, attend to the patient's emotional state or sensitivity.

The distinction, however, is not clear-cut, as all these therapies involve intense exploration of the patient's conflicts, and most rely on the emotion generated in therapy as a force in the patient's recovery.

In contrast, behaviour therapies derive from the view that neurosis is a matter of maladaptive conditioning and concentrate on modifying patients' behaviour.

There are arguments about the effectiveness of the various types of psychotherapy and art therapy, but it is generally agreed that success depends on a secure, confiding relationship between the therapist and client and on a shared confidence in the capacity of the therapist and his or her theory to explain and eliminate the problem.

END

ART FOR PSYCHOTHERAPY

INDEX	PAGE
Aetiology of the Disturbances	194
Aetiology	189
Aging	325
Amnesia	323
Ancient Mediterranean world	338
Ancillary Functions	224
Animals	151
Antithesis	194
Apparatus	197
Applied drawings	68
Arbitrary Functioning	229
Archetypes	236
ART FORMS FOR DIADNOSTIC METHODS USED	26
ART THERAPY & CONCEPTS	293
Artistic architectural drawings	79
Ascending Complex	230
Association Method	232
Attractions	187
Augustine's Term	236
Autobiographical memory	316
Autonomous Unconscious	228
Axial System	237
Bipolar Structure	236

ART FOR PSYCHOTHERAPY

Bodily Activity	**188**
Breaking up	**167**
Broadening of Consciousness	**229**
Brush drawings	**59**
Brush, pen, and dyestuffs	**51**
Camera supports	**275**
Carving tools and techniques	**123**
Carving	**121**
Casting and moulding	**131**
Cathexes	**193**
Cathexes and Hyper-cathexes	**199**
Causes of Amnesia	**327**
Centre of Reference	**211**
Centuries: 14th, 15th, and 16th	**80**
Centuries: 17th, 18th, and 19th	**84**
Chalks	**42**
Challenges to interference theory	**322**
Character Index	**222**
Charcoal	**40**
Child Development	**246**
Child Psychology	**248**
CINEMATOGRAPHY &TECHNOLOGY	**251**
Clarification	**237**
Coloured crayons	**50**
Combinations of various techniques	**62**

ART FOR PSYCHOTHERAPY

Comparative religions	353
Compass	225
Compensatory Relation	224
Complex	169
Complexes	231
Compromise	203
Conceptual history	337
Conclusion on magic and sorcery	356
Condensed Picture	207
Conflict and Rebellion	200
Conscious Adjustment	236
Conscious and Subconscious	163
Conscious Effort	223
Conscious Events	197
Conscious Processes	196
Conscious, Ego, Sub-conscious	166
Conscious, Preconscious, Ego, and Subconscious	173
Consciousness	195
Consciousness Assumption	195
Consciousness Dependent on Unconscious	213
Consciousness	211
Constructing and assembling	128
Continuity of Processes	234
Copernicus of the Mind	164
Creative Union	240

ART FOR PSYCHOTHERAPY

Critical of Freud	206
Culture of the Future	243
Decisive Facts	216
Decorative sculpture	155
Definitional issues: magic, religion, and science	333
Dependence	175
Development of the Sexual Function	185
Devotional images and narrative sculpture	149
Didactic and Diagnostic Method	233
Direct metal sculpture	130
Displacement	178
Disturbed Reality	205
Diversion	182
Doctrines	170
Dominant Function	217
Doodles	26
Double-system recording	288
Drawing surface	35
Drawing	28
Dream Interpretation	200
Dream Manifestation	201
Dream	233
Dream-process	201
Dreams Normality	200
Dreams	201

ART FOR PSYCHOTHERAPY

Dynamics and Economics	193
Eastern	90
Ego Forcing	201
Ego	174
Ego	210
Eight Different Psychological Types	225
Electra Phase	191
Electroplating	138
Elements and principles of design	30
Elements and principles of sculptural design	95
Elements of design	96
Elements of magic	329
Emotional Shock	232
Empirical Material	223
Energy	199
Equipment	242
Eros	179
Eros	184
Erotogenic Zones	184
Ethical approach	359
European traditions and the modern world	344
Evidence of Dream-work	202
Executive attention	301
Extraversion and Introversion	222
Eyewitness memory	317

ART FOR PSYCHOTHERAPY

Fanciful and non-representational drawings	78
Fantasies and Visions	235
Fantasy	152
Figure compositions and still lifes	77
Film processing and printing	282
Film	276
Fixation	182
Forgetting	318
Formulation of a session	359
Freud, Sigmund (1856-1939),	294
Fruitful Relationship	186
Function-complex	220
Functions	332
Fundamental Hypothesis	196
General characteristics of modelled sculpture	127
General methods	119
Gestalt	237
Gilding	137
Globalization of the magic concept	346
Graphite point	48
Habitus, the Central Switchboard	222
Heavy Burden	239
Historical Event	217
History of drawing – Western	80
History of magic theories	350

ART FOR PSYCHOTHERAPY

History	**251**
Hitler	**167**
Homosexuality	**192**
Human figure	**148**
Hypnosis	**164**
Hypotheses	**172**
Id	**174**
Id, Super-ego, and Reality	**176**
Impulses	**165**
Incised drawing	**50**
Independent Pleasure	**191**
Indirect carving	**122**
Individual Development	**172**
Infantile Amnesia	**188**
Influences on Sciences	**207**
Inhibitions	**192**
Inks	**53**
Inner Perception	**217**
Inner Rebuilding	**224**
Interchangeable Subjects	**209**
Interference	**319**
Internal Conflicts	**183**
Internal Resistances	**198**
Interpretation	**233**

ART FOR PSYCHOTHERAPY

Interpreting Freudian Concepts	163
Inter-relationship of Sciences	168
Intervals	*306*
Introduction of colour	260
Introduction of sound	256
Irrational Functions	216
Jung and Adler	167
Jungian Syntheses	205
Jungian System: Neither Religion nor Philosophy	242
Kingdom of the Illogical	203
Landscapes	74
Late medieval and early modern Europe	342
Libidinal Cathexes	183
Libidinal Demands	189
Libido	181
Light measurement	280
Light sources	277
Lighting	277
Listening	169
London Work	170
Long-term memory	304
Magic and religion	334
Magic and science	335
Magic	328
Magnetic recording	287

ART FOR PSYCHOTHERAPY

Manifold Contents	234
Material	330
Materials	105
Matrimonial Problems	224
Mechanical devices	66
Mechanism of Dreams	202
Medieval Europe	340
Memory	293
Mental Life	195
Mental Processes	198
Mental Regions	173
Mental Structure	195
Metalpoints	47
Methodology	170
Methods and techniques	116
Microphones	290
Mixed Types	218
Mnemonic systems	*307*
Modelling for casting	125
Modelling for pottery sculpture	127
Modelling	124
Modern forms of sculpture	144
Modern	89
Moral Conflict	231
Moral Standards	249

ART FOR PSYCHOTHERAPY

Motion pictures	**251**
Narcissistic	**183**
Natural Science	**197**
Nature and scope	**328**
Needs of the Id	**178**
Newer techniques	**291**
Non-representational sculpture	**153**
Nuclear Element	**229**
Observations and Experiences	**171**
Obstinate Persistence	**189**
Oedipus	**171**
Oedipus Phase	**190**
Opposing Forces	**180**
Optical recording	**286**
Orientation of Value	**223**
Other finishes	139
Other subjects	**152**
Over-differentiation	**219**
Painting	**137**
Passive State	**231**
Patination	**138**
Patterns of acquisition in long-term memory	**305**
Patterns of acquisition in working memory	**302**
Pen drawings	**55**
Penis Presence	**190**

ART FOR PSYCHOTHERAPY

Pens	**52**
Perceptions	**198**
Peremptory Influence	**199**
Persona	**219**
Personal Unconscious	**212**
Personality Inflation	**221**
Phallic Phase	**190**
Phenomena	**188**
Phenomenology	**193**
Philosophical Derivation	**208**
Philosophical Thought	**171**
Physical Apparatus	**176**
Physiological aspects of long-term memory	**309**
Plane techniques	**34**
Pointing	**135**
Portraits	**72**
Portraiture	**150**
Post-colonial points of views	**350**
Post-modern dialogue	**354**
Primal Datum	**212**
Primary Process	**199**
Primary	**105**
Principal parts	**268**
Principles of design	**99**
Principles	**210**

ART FOR PSYCHOTHERAPY

Product of Conflict	204
PROFESSIONAL ART THERAPY	357
Professional motion-picture production	268
Prophetic	235
Psyche Mechanism	232
Psychê	181
Psyche, Soul, or Mind	210
Psychiatry	244
Psychic Equals Physical	208
Psychic Functions	215
Psychic Health	221
Psychic Points	232
Psychic System	241
Psycho-analysis and Philosophy Disputes	197
Psychoanalysis in the Sphere of Philosophy	171
Psychoanalytical Society	166
Psychological Aspect	209
Psychological theories	352
Psychotherapies	243
Psychotropic drug	295
Puberty	192
Rational Functions	216
Reason over Instinctive Nature	243
Reassurance and Comfort	242
Region of Complex	177

ART FOR PSYCHOTHERAPY

Regional Divisions	228
Rehearsal	*307*
Relationship between drawing and other art forms	36
Relationships to other arts	104
Relearning	315
Relief sculpture	142
Remembering	297
Representational sculpture	147
Repression	165
Repressions	165
Reproduction and surface-finishing techniques	131
Restrictions	180
Retrieval	312
Rites and condition of the performer	331
Royal Pathway	233
Royal Society	168
Sadistic Impulses	189
Scenes of everyday life	150
Schematic comparisons	162
Science not Philosophy	208
SCRIBBLE, DOODLE, SKETCH, SCULPTURE	18
Sculptor as designer and as craftsman	117
Sculpture as an art	92
Sculpture in the round	139
SCULPTURE	92

ART FOR PSYCHOTHERAPY

Secondary	114
SECTION EIGHT: SORCERY OR MAGIC	328
Self –regulating System	206
Self-destructiveness	182
Self-preservation	179
Sexual Aggressiveness	180
Sexual Impulse	184
Sexuality	187
Short Task	207
Simultaneous Presence	191
Sketch	26
SKETCHES & DRAWINGS CONCEPTS	162
Skilled communication for therapy	357
Smoothing and polishing	136
Sociological theories	351
Somatic Processes	196
Sorcery	355
Sound recordist	289
Sound-recording techniques	256
Sphere of Consciousness	219
Sphere of Unconscious	227
Standard Symbols	234
State of consciousness	358
Stigmatisation	221
Studies and Experience	206

ART FOR PSYCHOTHERAPY

Subcategories of magic	336
Subconscious Mechanism	204
Subconscious Work-over	203
Subject matter of drawing	71
Substitutes	176
Super-ego Claims	205
Super-ego, Ego, Instincts	185
Superfluity	202
Surface finishing	136
Surfaces	37
Symbolism of sculpture	139
Symbolism	156
Symbolism	207
Symptom and Complex	229
Tendency	193
Tensions	175
Term Established	208
Theory of the Instincts	178
Time-dependent aspects of memory	300
Tools and techniques	39
Tranquillizer	297
Transformation	243
Typology	217
Unconscious Contents	214
Unconscious Sphere	213

ART FOR PSYCHOTHERAPY

UNDERSTANDING CHILDREN'S ART	**20**
Universal Human History	**235**
Unknown Points	**172**
Unsuccessful Adjustment	**222**
Uses of sculpture	**159**
Wide-screen and stereoscopic pictures	**265**
Wish Fulfilment	**204**
Working memory	**300**
World cultures	**347**
Writings	**168**
Zones	**228**

ART FOR PSYCHOTHERAPY

BIBLIOGRAPHY

1. AN OUTLINE OF PSYCHO-ANALYSIS. SIGMUND FREUD, THE HOGARTH PRESS, 1939

2. PSYCHO-ANALYSIS TRANSLATION. BY JAMES STRACHEY, 1949

3. THE INTERNATIONAL PSYCHO-ANALYTICAL LIBRARY. EDITED BY ERNEST JONES, 1949

4. SOME ELEMENTARY LESSONS IN PSYCHO-ANALYSIS. SIGMUND FREUD, 1938

5. VOLUME V OF FREUD'S COLLECTED PAPERS. SIGMUND FREUD, 1939

6. INTERNATIONAL JOURNAL OF PSYCHO-ANALYSIS, XXI. SIGMUND FREUD, 1940

7. THE PSYCHO-ANALYSIS OF CHILDREN. ANNA FREUD, 1942

8. CHILDREN'S PSYCHOANALYSIS. MELANIE KLEIN, 1949

9. DIE PSYCHOLOGIE VON C G JUNG. K W BASH, 1942

10. THE PSYCHOLOGY OF C G JUNG. JOLAN JACOBI, KEGAN PAUL, TRENCH, TRUBNER & CO LTD 1943

11. PSYCHOLOGY OF THE UNCONSCIOUS. C G JUNG, 1913

12. INDIVIDUAL PSYCHOLOGY. ALFRED ADLER, 1914

13. TOTEM AND TABU. SIGMUND FREUD

14. STUDIEN ÜBER HYSTERIE. BREUER AND FREUD

ART FOR PSYCHOTHERAPY

15. BEYOND THE PLEASURE PRINCIPLE. SIGMUND FREUD

16. EGO AND I. SIGMUND FREUD

17. THE FUTURE OF AN ILLUSION... SIGMUND FREUD

18. WHY WAR? SIGMUND FREUD AND ALBERT EINSTEIN, 1933.

19. THE PSYCHOPATHOLOGY OF EVERYDAY LIFE. SIGMUND FREUD, 1938

20. THE INTERPRETATION OF DREAMS. SIGMUND FREUD, 1936.

ALL PUBLICATIONS MENTIONED BELOW ARE WRITTEN BY ANDREAS SOFRONIOU:

21. MEDICAL ETHICS THROUGH THE AGES, ISBN: 978-1-4092- 7468-1

22. THE MISINTERPRETATION OF SIGMUND FREUD, ISBN: 978-1-4467-1659-5

23. JUNG'S PSYCHOTHERAPY: PSYCHOLOGICAL & MYTHOLOGICAL METHODS, ISBN: 978-1-4477-4740-6

24. FREUDIAN ANALYSIS & JUNGIAN SYNTHESIS, ISBN: 978-1-4477-5996-6

25. PSYCHOTHERAPY, CONCEPTS OF TREATMENT, ISBN: 978-1-291-50178-0

26. PSYCHOLOGY, CONCEPTS OF BEHAVIOUR, ISBN: 978-1-291-47573-9

27. PHILOSOPHY FOR HUMAN BEHAVIOUR, ISBN: 978-1-291-12707-2

ART FOR PSYCHOTHERAPY

28. SEX, AN EXPLORATION OF SEXUALITY, EROS AND LOVE, ISBN: 978-1-291-56931-5

29. PSYCHOLOGY OF CHILD CULTURE, ISBN: 978-1-4092-7619-7

30. THE GUIDE TO A JOYFUL PARENTING, ISBN: 0 952 7956 1 2

31. THERAPEUTIC PHILOSOPHY FOR THE INDIVIDUAL AND THE STATE, ISBN: 978-1-4092-7586-2

32. PHILOSOPHIC COUNSELLING FOR PEOPLE AND THEIR GOVERNMENTS, ISBN: 978-1-4092-7400-1

33. SOCIOLOGY, CONCEPTS OF GROUP BEHAVIOUR, ISBN: 978-1-291-51888-7

34. SOCIAL SCIENCES, CONCEPTS OF BRANCHES AND RELATIONSHIPS ISBN: 978-1-291-52321-8

35. CONCEPTS OF SOCIAL SCIENTISTS AND GREAT THINKERS, ISBN: 978-1-291-53786-4.

36. MORAL PHILOSOPHY, THE ETHICAL APPROACH THROUGH THE AGES, ISBN: 978-1-4092-7703-3

37. 2011 POLITICS, ORGANISATIONS, PSYCHOANALYSIS, POETRY, ISBN: 978-1-4467-2741-6

38. SOCIAL SCIENCES AND PHILOLOGY, ISBN: 978-1-326-33840-4

39. PHILOLOGY, CONCEPTS OF EUROPEAN LITERATURE, ISBN: 978-1-291-49148-7

40. SOCIOLOGY, CONCEPTS OF GROUP BEHAVIOUR, ISBN:

ART FOR PSYCHOTHERAPY

978-1-291-51888-7

41. SOCIAL SCIENCES, CONCEPTS OF BRANCHES AND RELATIONSHIPS ISBN: 978-1-291-52321-8

42. CONCEPTS OF SOCIAL SCIENTISTS AND GREAT THINKERS, ISBN: 978-1-291-53786-4

43. THERAPEUTIC PSYCHOLOGY, ISBN: 978-1-326-34523-5

44. MEDICAL ETHICS THROUGH THE AGES, ISBN: 978-1-4092-7468-1

45. MISINTERPRETATION OF SIGMUND FREUD, ISBN: 978-1-4467-1659-5

46. JUNG'S PSYCHOTHERAPY: THE PSYCHOLOGICAL & MYTHOLOGICAL METHODS, ISBN: 978-1-4477-4740-6

47. FREUDIAN ANALYSIS & JUNGIAN SYNTHESIS, ISBN: 978-1-4477-5996-6

48. ADLER'S INDIVIDUAL PSYCHOLOGY AND RELATED METHODS, ISBN: 978-1-291-85951-5

49. ADLERIAN INDIVIDUALISM , JUNGIAN SYNTHESIS, FREUDIAN ANALYSIS, ISBN: 978-1-291-85937-9

50. PSYCHOTHERAPY, CONCEPTS OF TREATMENT, ISBN: 978-1-291-50178-0

51. PSYCHOLOGY, CONCEPTS OF BEHAVIOUR, ISBN: 978-1-291-47573-9

52. PHILOSOPHY FOR HUMAN BEHAVIOUR, ISBN: 978-1-291-12707-2

53. SEX, AN EXPLORATION OF SEXUALITY, EROS AND

ART FOR PSYCHOTHERAPY

LOVE, ISBN: 978-1-291-56931-5

54. PSYCHOLOGY FROM CONCEPTION TO SENILITY, ISBN: 978-1-4092-7218-2

55. PSYCHOLOGY OF CHILD CULTURE, ISBN: 978-1-4092-7619-7

56. JOYFUL PARENTING, ISBN: 0 9527956 1 2

57. GUIDE TO A JOYFUL PARENTING, ISBN: 0 952 7956 1 2

58. THERAPEUTIC PHILOSOPHY FOR THE INDIVIDUAL AND THE STATE, ISBN: 978-1-4092-7586-2

59. CHILD PSYCHOTHERAPY, ISBN: 978-1-326-44169-2.

60. THERAPEUTIC PHILOSOPHY FOR THE INDIVIDUAL AND THE STATE, ISBN: 978-1-4092-7586-2

61. PHILOSOPHIC COUNSELLING FOR PEOPLE AND THEIR GOVERNMENTS, ISBN: 978-1-4092-7400-1

62. MORAL PHILOSOPHY, THE ETHICAL APPROACH THROUGH THE AGES, ISBN: 978-1-4092-7703-3

63. PSYCHOANALYSIS, POETRY, ISBN: 978-1-4467-2741-6

64. PHILOLOGY, CONCEPTS OF EUROPEAN LITERATURE, ISBN: 978-1-291-49148-7

65. SOCIOLOGY, CONCEPTS OF GROUP BEHAVIOUR, ISBN: 978-1-291-51888-7

66. SOCIAL SCIENCES, CONCEPTS OF BRANCHES AND RELATIONSHIPS, ISBN: 978-1-291-52321-8

67. CONCEPTS OF SOCIAL SCIENTISTS AND GREAT

ART FOR PSYCHOTHERAPY

THINKERS, ISBN: 978-1-291-53786-4

68. MEDICAL ETHICS THROUGH THE AGES, ISBN: 978-1-4092-7468-1

69. MEDICAL ETHICS, FROM HIPPOCRATES TO THE 21ST CENTURY ISBN: 978-1-4457-1203-1

70. THE MISINTERPRETATION OF SIGMUND FREUD, ISBN: 978-1-4467-1659-5

71. JUNG'S PSYCHOTHERAPY: THE PSYCHOLOGICAL & MYTHOLOGICAL METHODS, ISBN: 978-1-4477-4740-6

72. FREUDIAN ANALYSIS & JUNGIAN SYNTHESIS, ISBN: 978-1-4477-5996-6

73. PSYCHOLOGY FROM CONCEPTION TO SENILITY, ISBN: 978-1-4092-7218-2

74. PSYCHOTHERAPY, CONCEPTS OF TREATMENT, ISBN: 978-1-291-50178-0

75. PSYCHOLOGY, CONCEPTS OF BEHAVIOUR, ISBN: 978-1-291-47573-9

76. PSYCHOLOGY OF CHILD CULTURE, ISBN: 978-1-4092-7619-7

77. THE GUIDE TO A JOYFUL PARENTING, ISBN: 0 952 7956 1 2

78. PHILOSOPHY FOR HUMAN BEHAVIOUR, ISBN: 978-1-291-12707-2

www.ingramcontent.com/pod-product-compliance
Lightning Source LLC
Chambersburg PA
CBHW060820170526
45158CB00001B/33